BUSINESS IN JAPAN

Business in Japan

*A guide to
Japanese business
practice and
procedure*

REVISED EDITION

*Edited by Paul Norbury
and Geoffrey Bownas*

M

First edition 1974
Second edition (fully revised) 1980

First published in 1974 by
THE MACMILLAN PRESS LTD

Second edition 1980 published by
THE MACMILLAN PRESS LTD
London and Basingstoke
Companies and representatives throughout the world
and
PAUL NORBURY PUBLICATIONS LTD
Tenterden, Kent

*Photoset and printed by Bookmag, Inverness and bound by Hunter
& Foulis, Edinburgh*

British Library Cataloguing in Publication Data

Business in Japan. 2nd ed.
  1. Business enterprises — Japan
  2. Japan — commerce
  I. Norbury, Paul II. Bownas, Geoffrey
  338.7'C 0952 HF3826.5

ISBN 0 333 30001 7

It all started thirty years ago when I first flew from my native Brussels to Tokyo, stopping along the way for a few weeks in Calcutta and Shanghai. On my second day in Tokyo, I climbed into a streetcar that obviously had survived the war at great cost. As my eyes wandered over the strange people and things surrounding me, my bewilderment grew. Wasn't there somewhere something familiar my eyes could cling to? Suddenly, in a corner near the ceiling of the streetcar, I recognised familiar symbols. Something was written in English. At closer scrutiny, it read 'Made in Japan'. In Calcutta and Shanghai, what I had seen of the East was its misery, and whenever I saw a familiar convenience, it was always made in England, in France, or in the U.S.A. The 'Made in Japan' mark I saw in the streetcar, however, carried a different message: Japan was and yet was not the East. Of course, the ready explanation was that the Japanese were imitating us; but I could not answer the question of why the Indians and the Chinese had not done the same.

I was soon to discover that many other seemingly familiar things were also 'Made in Japan'; for example, eggs. In those early post-war years, the taste of eggs in Japan was that of feed given to chickens — fish. Eggs looked perfectly familiar, but their taste was not.

At any rate, the words 'Made in Japan' were to haunt me for the following three decades. And in trying to understand the phenomenon behind those words, I found little help in visiting the traditional old temples, where I would join throngs of Japanese who were likewise 'visiting' their past. The challenge was not in the past but in the present.

ROBERT BALLON

In 1972 Japan Air Lines introduced the JAL EXECUTIVE SERVICE to aid the international business traveller visiting Japan and the Far East. As part of this service a series of booklets was produced by JAL entitled *Business in Japan — Guidelines for Exporters*. These booklets were brought together to form part of the first edition of *Business in Japan* published at the end of 1974. This new edition contains a considerable amount of new material as well as some of the original chapters now fully revised.

The Japanese, because of their history and culture, have a different approach to commerce compared to the rest of the world. Communication and understanding are important in any venture but in the field of business relationships they are paramount. It is our sincere hope that this book will help to provide an insight to the ways and customs of Japanese business procedure and that it will assist in fostering trade relations with Japan in an ever expanding commercial world.

JAPAN AIR LINES

Contents

Foreword   ix
Editors' Preface   xi
Glossary of Japanese terms used in the text   xiii

SECTION I   UNDERSTANDING THE JAPANESE
  1   The Logic Gap *Martyn Naylor*   1
  2   Japan's Unique Group Dynamic *Gregory Clark*   5

*SECTION II   THE NEW INDUSTRIAL POLICIES OF
              JAPAN*
  3   Japanese Industries to 1985 *Shinzo Katada*   10
  4   Japan and the New Industrial Countries of East Asia   21
      Charles Smith

SECTION III APPROACHES TO THE MARKET
  5   Working with Japan's Free Market Structure *Gene*   29
      Gregory
  6   Pointers to Success and Failure *Sadao Oba*   36
  7   Japan's New Superconsumers *Teruyasu Murakami*   40
  8   Rationalising the Distribution System *Masao Okamoto*   57
  9   The Value of Market Research *Andrew Watt*   63
 10   The Role and Application of Advertising *David Gribbin*   68
 11   The Japanese Housewife: a Marketing Appraisal *George*   76
      Fields
 12   Marketing Tailpiece: the 'Tranny' and the Fridge   82
      Masaaki Imai

SECTION IV FINANCE AND THE BANKS
 13   The Japanese and their Changing Economic   85
      Environment *John Kirby*
 14   Foreign Banks in Japan *John Robinson*   94
 15   Services of a Japanese Bank *Kunihiko Kobayashi &*   98
      Takashi Sugiyama
 16   Stocks, Securities and the Brokerage Market *Simon*   102
      Grove

SECTION V STRATEGY & MANAGEMENT
 17   Aspects of Japan's business interrelationships *Michael*   109
      Isherwood
 18   Management Style *Robert Ballon*   115

19 Japanese Trading Companies in Transition *Sadao Oba* 131
20 Joint Ventures *Robert Ballon* 137
21 Business and the Law *Sumio Takeuchi* 155

SECTION VI ADJUSTING TO JAPAN
22 Setting up a Small Office *Simon Grove* 163
23 Smiths Industries: A European Case-Study *Jonathan* 173
   Rice
24 Etiquette and Behaviour *George Fields* 177
25 Check your own Check-list *Michael Isherwood* 183
Epilogue Reflections on Relationships *Helmut Morsbach* 190

Bibliography 196
List of Contributors 203
Anatomy of a Japanese trading company (Diagram) 206
Index 207

Foreword

Japan represents not only the largest industrial and consumer market in Asia, it is the third largest market in the world. Although it was often liberalised to foreign imports tardily and apparently somewhat reluctantly, it is now, by and large, an open market. There are, it is true, major changes in economic structure which are needed before Japan's imports of manufactured goods are increased to bring them closer to the same relative levels as those of the European Community or of the United States. It will take time, and efforts on both sides, before the economies of these three principal partners are more fully integrated with one another. But individual European companies that have persevered have done excellent business in Japan (EC exports to Japan have increased five-fold between 1968 and 1979). They have also helped maintain their position in third markets by preventing potential Japanese rivals from gaining strength on their own home ground.

Japan's successful export drives to the European market have been based on a lot of hard-won knowledge of Europe and of European industrial and business practices. It goes without saying that Europeans, or indeed any exporter, cannot hope to succeed in the Japanese market without a similar sustained commitment to learn about Japan. I can think of no better starting point than *Business in Japan* which contains both lucid analyses of Japan's 'business culture' and much shrewd advice on how to enter this difficult but rewarding market.

ENDYMION WILKINSON

Endymion Wilkinson spent six years in Japan (1974–9) as head of the economic and commercial department of the EEC's Delegation in Tokyo. He is currently at the China desk of the EEC Commission, Brussels.

Editors' Preface

In this second edition of *Business in Japan* we have endeavoured to cover as before the main practical elements of doing business with the Japanese. There are many differences and there are some similarities; above all, however, there are countless nuances and subtleties within the framework of what we could call Japan's business culture. Whatever the dos and don'ts may be, whatever the specific points of reference in law, in finance, in joint ventures, in distribution and in trading generally may be, the businessman will be thwarting his own long-term future associations if he does not approach Japan on a broad socio-cultural basis. For the deeper our grasp and understanding of Japanese society, the better are we going to be served and serve ourselves in our dealings with Japan.

Hence the first section entitled 'Understanding the Japanese'. Through the varied views and interpretations put forward the message is: search for both an intellectual and emotional grasp of who the Japanese are. But remember that however close you come, you are dealing with the world's greatest ethnic monoculture that Gregory Clark likens to the original 'particularistic' tribal societies we all once were — exclusive to outside people but not to outside ideas.

Another point to bear in mind is the considerable and, at times, devastating element of surprise and apparent contradiction which pervades Japanese society (as perceived by the outsider). It is often said that after two weeks in a strange country one can write a book about it. After two months things seem a little more complicated. After two years we realise that we are faced with a life-long study and even then the real level of understanding might not be much more than surface deep. The more familiar we become with Japan the less able, it seems, we are to predict or to generalise; the more we observe and experience this 'particularistic' society, the more we become aware (and wary) of what are essentially alien thought and emotional responses. Martyn Naylor, the other contributor in the first section, calls this the 'logic gap'.

This new edition of *Business in Japan*, therefore, could be said to have grown up somewhat since it was first published in late 1974. In a very real sense the world as a whole has grown up, too, at least in economic terms: it has had to face the harsh realities of a withering recession after the 1973 oil crisis and finite resources.

Japan, the most resource-dependent of all the world's major economies, harboured as it is alongside the troubled continent of Asia, has come to terms more successfully than anyone else with all of these limitations and at the same time has maintained the national equilibrium and status quo. While most of the world shuffled its governments around during this period in an attempt to bring about a new order and new priorities, Japan kept a steady course which if anything is today even steadier; and with the eighties about to begin we are likely to see an even greater consolidation of her position.

Like her printed integrated circuits, heralding the new computer age, Japan's integrated social structure will be seen increasingly as her single greatest asset. Harmony is not just a social convention in Japan, it is a need and, as such, gives her the capacity to exploit any opportunity or situation to the full. Indeed, the people of Japan have always been her great micro-chip resource since the dawn of history. Similarly, we feel that Europe, the United States and the western business world as a whole has grown up in its attitude to Japan. One is no longer preaching that the Japanese market actually exists (which was the case from the beginning of the sixties to the early seventies); today we can discuss *how* to penetrate and exploit it. Britain, for example, doubled her exports to Japan in the five years 1973-78, and in FY78-79 Japan's total imports were greater than ever before.

The approach to this edition, therefore, has been practical, pragmatic and, we hope, even more perceptive. We have tried to make available more value-judgements based on expertise and hard-earned experience. In our Preface to the first edition, written as Japan faced the 'oil shock', we wrote of our confidence in her ability to ride out the storm. Now, as the eighties dawn, we feel just as convinced that the decade will belong to Japan — whether as importer, exporter, joint venturer, overseas investor or fertile technological leader. What is more, as a key member of the Pacific Region, we have no doubt that Japan will play an increasingly critical role in the developments that take place.

Through the pages of this book, we hope we have given the reader sufficient background to understand and associate with the Japan of the 1980s as well as support him with the know-how for searching out and grasping the rich opportunities that lie ahead.

20 NOVEMBER 1979

Paul Norbury
Geoffrey Bownas

Glossary

Agaru	Literally 'to step up' into a Japanese house, from the *genkan* (see below): to enter a home, belong to a family	*JETRO*	Japan External Trade Organization
		Jomukai	Executive committee
		Ka	Section
Amae	See chapter 2, p.5 (impossible to give simple definition)	*Kabuki*	A dramatic form, originating in the 17th century
		Kabushiki kaisha	Joint-stock corporation
Amakudari	Descending from heaven = upon voluntary retirement from the Civil Service move into a directorship in some large corporation	*Kacho*	Section chief
		Kafu	Family style
		Kaicho	Chairman of board of directors
		Kana	Japanese syllabary
Asobi	Play	*Kanji*	Sino-Chinese character
Bu	Department	*Kansa hojin*	Auditing corporation
Bucho	Department head	*Kansa yaku*	Statutory auditor
Bun	Status, role	*Kato kyoso*	Excessive competition
Daimyo	Feudal lord	*Kawaigaru*	To pet, to favour a subordinate
Denden Kosha	The Telegraph and Telephone Corporation (NTT)	*Keidanren*	Japan Federation of Economic Organizations
		Keizai doyukai	Japan Committee for Economic Development
Fu	Style		
Futon	Mattress	*Kenrikin*	Key money (see also *reikin*)
Gaijin	Foreigner		
Geisha	Hostess, literally 'a person of the arts'	*Kobun*	Followers
		Kogaisha	'Child company'
Genkan	Doorway, entrance (the boundary between home and the outside world)	*Kokusai Denwa Denshin Kyoku*	(KDD) Overseas Telephone/Telecommunications Bureau
Gensaki	On the stock exchange: spot/future		
		Kunigara	National characteristics
Giri	Duty	*Kyoso*	Competition
Giri-ninjo	(The conflict of) duty and emotion or human feeling	*Manga*	Cartoon, comic
		Meishi	Name card
Gyosei shido	Administrative guidance	*Nijikai*	Second get together
Haiku	Seventeen-syllable poem, typical of 17th century literature	*No*	Traditional dramatic form, originating in 14th century
Hanko	Seal	*On*	Obligation
Haragei	Stomach art, gut feeling	*Reikin*	Caution money (see also *kenrikin*)
Hashigozake	Drinking up the (entertainment) ladder (see *Nijikai*)		
		Ringisho	Circulation of a document
		Rirekisho	Curriculum vitae
Honne	Real intention	*Saitori*	Stock exchange clerk
Hoshokin	Security, advance rent (see also *shikikin*)	*Samurai*	Knight, warrior
		Seishin	Training of one's own inner mental strength
Ichijikai	First get together		
Iemoto	Literally 'family root': the head of a school of the polite arts	*Sensei*	Teacher
		Shacho	Company president
		Shafu	Company style

Shikikin	Deposit, caution money (see *hoshokin*)	Tatemae	See Chapter 9, p.
		Tofu	Bean curd
Shingikai	Joint consultative committees	Torishi-mariyaku	Director
Shogun	Military-administrative leader of Japan, 12th to 19th centuries	Tsubo	3.3 square metres
		Wakon yosai	'Japanese spirit, western material' — the eclectic philosophy of the Meiji period reformers
Shokuba	Work place		
Shugaku-ryoko	Study tour, school trip		
		Yugen kaisha	Limited liability company
Shukko	Transfer	Zaibatsu	Pre-war large group of companies, still used today to describe a type of operation
Sogo shosha	Trading corporation		
Soroban	Abacus		
Sumo	Traditional wrestling style		
Taishokkin	Termination bonus		

1
The Logic Gap

MARTYN NAYLOR

Japan is deceptively like any other market in the industrialised world, and it is all too easy to be impressed by the similarities and to minimise, or even disregard, the differences. However, a basic problem is that Japan is a land of paradoxes, and when discussing Japan it is often possible to present both sides of an argument with equal conviction — the Japanese are the most polite *and* the rudest people on earth; most Japanese tend to appreciate beauty more readily than their western counterparts, yet they seem content to accept the urban clutter and confusion of their largely characterless major cities; the Japanese economy and infrastructure are highly developed, yet little more than one-third of all homes has a flush toilet (and housing is generally poor by the standards of other industrialised countries); the Japanese are highly sensitive and care what people think about them, both individually and as a nation, yet as a member of the world community they often appear to be shirking their responsibilities. Priorities are different from those of the West in many instances, whilst thought processes often follow a parallel plane.

Living and working in Japan usually requires a higher degree of adaptation and re-adjustment for the foreigner than in other developed countries, basically as a direct result of this apparent dual personality of the Japanese. Whilst the foreigner is welcomed and generally accepted on the one hand, there are continual reminders that he is alien to Japanese society on the other.

Perhaps the key to success in Japan, basic even to survival itself, is to refrain from 'rocking the boat'. This does not mean to say that one should adopt a 'low profile' as is so often done by Japan as regards other nations; one can be sufficiently aggressive to tackle any task competitively, but this must be accomplished within the limitations set by the Japanese 'system'. By all means devise methods of avoiding these restrictions, established by law or convention, but under no circumstances challenge the system or engage in a head-on confrontation.

The root of so many misunderstandings on the part of the foreign businessman in Japan is what can best be described, perhaps, as the *logic gap* — a communication barrier resulting not so much from language as a set of values and thought processes that, if one can generalise, run parallel with the West's and never meet. Whilst this in large part developed as a direct result of introspection conditioned by Japan's centuries of isolation, it tends to be sustained by the group conditions inherent in society today. It could be argued that the *logic gap* is merely a convenient way for the apologist to rationalise the Japanese inability to see the other person's point of view, and the reason why an educated person can often appear, in western terms, to be expecting both to have his cake and eat it. But the 'uniqueness' of this set of rules the Japanese appear to be living by depends to a great extent on how we ourselves assess and interpret motives.

There is the story of the sportsman who came to Japan, was first past the post in his race, and was dismayed when he was offered no place on the winner's rostrum; he could hardly be expected to be satisfied with the explanation of the Japanese organisers that he was only invited to participate in the race, not to win it! Such apparent duplicity is clearly unconscious, although it is difficult for the newcomer to feel that it is anything but an overt strategy designed to thwart his competitive efforts.

Extended to the international market-place at a time when Japan has a severe trade imbalance with Western Europe and North America, it appears to the outside observer that Japan is disinclined to take action to redress the current imbalance, and he sees it as indefensible that she should be unwilling to accept the responsibilities of a mature economic power. It is suggested that the Japanese market should be further opened to foreign products, but it is difficult for the Japanese, businessman and bureaucrat alike, to appreciate many of the apparent restrictions that remain. However, whilst the resident foreign businessman can readily identify a range of non-tariff barriers, these must be recognised by the Japanese before they can be dismantled; there are severe limits to the extent to which a bureaucrat will question the regulations he is required to administer, and practical considerations dictate, for example, the preparation of a positive rather than a negative list for items to be carefully controlled. All products, regardless of whether they are imported or produced domestically, must contend with the inefficiencies of the Japanese distribution system: this, in theory, leaves the field wide open; but, in practice, any imported product new to the market encounters enormous odds against gaining an initial toe-hold.

How does this affect the expatriate going about his routine tasks of doing business in Japan? First, it introduces a duality into his life that he may well be unprepared for. If he is reporting back to a head office overseas he may have to spend a disproportionate amount of time interpreting Japanese conditions to a disbelieving management at the home office, since actual experience may differ widely from what was discussed during visits to Japan. Then he must be able to deal with his own staff and associates on terms acceptable to them, and this invariably requires a degree of self-adjustment to the local situation in which he must consciously seek a role where he can work in harmony with the Japanese side rather than in conflict with it. This is precisely where so many problems develop, be this a relationship with own staff/subordinates, with joint venture partners, or with local distributors; it is a matter of two-way communication, which is rendered of minimal effectiveness not because of a language barrier, but as a result of parallel methods of thinking about the same subject — the *logic gap* mentioned earlier. The expatriate is in a minority position, so the ball is in his court and he is the one who should be expected to make the greatest degree of adjustment.

How, then, can adjustment be made? It is certainly much simpler than my reflections above might lead you to believe. The key is to understand that there may be alternative ways to tackle a familiar problem, that the usual rationale may be supplanted by another of equal validity, that a short-term compromise can avoid the embarrassment and inconvenience of a head-on collision, and that it is often essential to conform to the group point of view. In other words, don't rock the boat, don't stick your neck out too far or too quickly, don't try to provide all the answers at once, don't forget there is probably another viewpoint, and don't assert that experience elsewhere must be valid for Japan. In case this might sound like a plea for passivity, it should be understood that the root cause of the lack of success of many products in the Japanese market, not to mention the failure of a number of significant joint ventures, has been the undue aggressiveness and misguided judgement of local expatriate management, who have tried to impose their own ideas in sharp resistance to the prevailing system in Japan.

The Japanese generally view themselves as a homogeneous race, which is by no means correct, and they like to think they are different from others. By all means pander to these whims, since there is little to be gained by asserting that the Japanese are essentially the same as businessmen and consumers around the world, yet in this paradoxical society the similarities tend to be immediately more obvious. It is essential to work with Japanese colleagues and associates on an equal basis, yet this, again, is a two-way thing, and it is often the Japanese side which will not accept an expatriate, and leave him as no more than an outsider or figure-head in most discussions. To avoid such a predicament, try to be patient and show that you are prepared to make efforts to understand the situation in Japan, but be conciliatory only to the point where you do not lower your own standards, being alert to opportunities, especially those of exploiting any differences to real advantage.

You should also appreciate that the Japanese are usually highly sensitive and can be easily offended; yet there will be times when they can be completely insensitive, especially in relation to the problems you, personally, are facing.

Your Japanese associates will show you more respect if you do not attempt to change everything that fails to conform to business practice previously experienced, or challenge any regulations that seem illogical. If one route appears impossible, seek out alternatives that may circumvent the barriers.

You might feel that most of this could be said about almost any other nation on earth, and you would probably be right. Perhaps Japan is not so different after all, but in few other places would you have to work so hard to convince yourself that things are much the same as elsewhere. Certainly, Japan can be highly frustrating for the foreign businessman, but with a fair degree of patience, insight into the Japanese character, and marketing skills, not to mention general business acumen, this can become one of the most rewarding markets in the world.

2
Japan's Unique Group Dynamic

GREGORY CLARK

When I first came to Japan I was told that the Japanese were a subservient, disciplined people who lived in a rigidly hierarchical society. Shortly after that, I saw my first labour disputes — hundreds of younger workers waving red flags and blockading the main entrance to the company (a rather conservative newspaper group); insulting posters plastered over the president's door, demure secretaries with threatening armbands, activists with bull horns marching into offices. I realised that what I had been told about Japan might not be quite right.

There are many theories to explain the Japanese. Some emphasise the instinctive sensitivity to hierarchy (the vertical society). Some emphasise the instinctive urge to dependency relations (the amae society). Others emphasise one or more of the many other unusual attitudes or values found in Japan — group relations, giri-ninjo (obligation-compassion) and so on. And the reasons why these qualities are said to be found uniquely in Japan range from rice culture and alleged racial homogeneity to island location and climate.

One of the latest theories says that the structure of the Japanese brain is different from that of all other people and this is the cause of the pure emotional vowel tones in the Japanese language. As a result the Japanese are an emotional people sensitive to nature and insect sounds but unable to operate in the abstract, logical terms that non-Japanese prefer.

My own theory starts from the assumption that all men and races are fundamentally the same. And we all carry two dimensions to our personality — the emotional, the instinctive, and the sensitivity to immediate human relations on the one hand, and on the other the ability to operate in terms of the intellectual, the principled and the scientific. Or as the sociologists put it, the particularistic and the universal. We use both as the situation demands. What makes the Japanese unusual is that they alone among the advanced peoples have clung to the particularistic as the basis of their value system.

As to why this has happened, I suggest it has something to do with Japan's relative lack of foreign wars and foreign enemies for most of its history. Such wars and enemies force nations to turn from particularistic to universal values as they seek to create the ideologies and national identity needed to resist or conquer the foreign enemy. Japan has not experienced this need, so it has simply remained with its original particularist, 'tribal' values and refined them to allow the organisation of a large, complex society. It does not have the same urge as the rest of us to seek for absolutes, for 'objective' principles and reasons and apply them even to the details of everyday life. It is quite happy to stay with the emotional and instinctive, and to rely on unargued custom and tradition for its rules.

But none of this denies the *ability*, as opposed to the *propensity*, of the Japanese to operate in terms of the universalistic. They have, as the need required, introduced science and ideologies, the concept of law, contracts and central government — all universalistic features. But they do not actively seek out these things. They have intelligence, but not intellectuality. In a situation where there is a choice between an abstract, argued approach and one that appeals to the human, emotional sensitivities they will prefer the latter. Those who prefer the former are criticised as *rikutsupoi*, or too inclined to operate in terms of reasons.

The concept of the Japanese as an 'emotional' people is often hard for westerners to understand. We associate the word with the open display of feelings and personality. We see the anti-individualism and constraints of Japanese society as un-emotional. But if we use the word in its deeper sense of constructing a value system on the basis of human relationships rather than principles and ideologies, the meaning will be more apparent. Precisely because the emotional is so important in Japan the open display of feelings has to be controlled lest it harm the delicate fabric of relationships.

Certainly the Japanese have no hesitation in using the word emotional to describe themselves. And those who know Japan realise that even in the western sense of the word it lingers close to the Japanese surface, to emerge in the violence of demonstrations, the uninhibited festivals and the intimacy of close relationships.

In the negative sense of being non-intellectual it is also apparent daily in everything from the refusal of newspapers and politicians to take clear stands on issues to the sloppy arguments used by academics to promote their pet theories.

All this immediately raises the question of how such a non-intellectual people could have organised one of the world's more sophisticated societies, not to mention a very powerful economy. And the answer, I suggest, requires us to revise our previous view about these things. Social and economic progress may not be entirely the result of a move to the universalistic approach. True, in the development of pure sciences or philosophies the more universalism the more progress. But in the development of societies and economies a balance is needed. The Japanese with their openness to the universalistic but preference for the particularistic may come closer than the rest of us to that balance.

Perhaps the best example of this is enterprise formation. To date we have assumed that a high degree of universalism is needed to persuade people to break away from the cloying ties with clan, caste, class or region and join cooperatively and productively in production activities. I would argue the reverse. Clan, caste, class or regional affiliations are all universalistic groups since they bring people together on the basis of abstract attributes. The attributes required to join the group — religious belief, place of birth and upbringing, and so on — are a 'principle', admittedly a very crude principle, on which people relate to each other. And the proof of this is the ability of people who have never met each other to relate to each other on the basis of this principle.

But the enterprise, like the village and the nuclear family, says that people brought together in the same location should relate to each other simply because of the accident of their being brought together. In the process they ignore regional, religious, class and other affiliations. Or rather they regard them as secondary. But this is not a proof that the Japanese have moved to a superior and more rational set of 'principles' on which to form groups. On the contrary. It means that unlike the rest of us they can turn to the emotional and particularistic as the basis of group formation.

The proof of this is the ease with which factions form within Japanese groups and the instinctive exclusiveness to those outside those groups. In other words the Japanese relate closely to those with whom they are in contact, but not to others. 'The distant relative is less important than the nearby stranger', is a Japanese saying. The abstract affiliation of blood is less binding than the fact of physical proximity. It is 'unprincipled', but highly effective in the organisation of enterprises.

This ability of the Japanese to come together easily into enterprise groups is a major strength for the Japanese economy. Admittedly, it is helped by management techniques that encourage instinctive, emotional relationships to and within the enterprise — paternalistic management, seniority promotion, life-time employment, staff facilities, consensus decisions and so on. But the main factor is the retention by Japan of particularistic values. In the more universalistic societies workers cling to a more principled basis of identification —

individualism, trade and professional groups for example — often with a destructive clash of loyalties as a result. And while our more 'principled' management techniques — merit promotion, hire and fire, top-down management — have rationality on their side they are often destructive of the emotional loyalties and links that could also be used to bind western workers to their workplace. In the highly universalistic societies such as China and India, most enterprise groups are unable to provide an alternative to the 'principled' ties of clan or caste.

The other major economic plus for the Japanese is eclecticism. All universalistic peoples tend automatically to reject the easy inflow of ideas and culture from outside. To allow such an inflow denies the validity of the ideas and culture — the intellectual basis — on which they seek their identity. So they need to be convinced that the new ideas are of benefit and have validity. Once so convinced they will take in firmly the outside ideas and make them their own.

The Japanese do not operate this way. Since they see identity in more emotional, human relations terms they do not feel threatened by outside 'intellectual' intrusions. So they are exclusive to outside people, but not to outside ideas. To universalistic man this may seem to be an 'unprincipled' approach but it is highly effective when it comes to promoting economic growth. It has a disadvantage in that the idea imported from outside is often not fully understood or integrated into the value system. But that is minor when compared with the pragmatic benefits that accrue.

In most other areas the disadvantages of the particularistic approach work more closely to cancel out the advantages. For example, the Japanese ability to concentrate on an immediate task or situation is balanced out by their relative weakness in long-term and abstract planning. They are good at tactics but not at strategy. An ability to mobilise and change course rapidly on the basis of emotional mood is balanced by an inability to do the same thing on the basis of argument and principle. (When they finally created the national consensus to tackle the pollution problem they did so very well. But this was of little consolation to those who suffered in the long years when anti-pollution measures were clearly needed, but the appropriate mood and consensus had not been created.) The flexibility of the non-legalistic approach is negated by some of the arbitrary injustices that result.

Even so the net merits of the particularistic approach should not be overlooked. The weakness in pure science is compensated greatly by the concentration in applied science. The lack of interest in ideology and issues does at least lead to considerable social stability. And the weaknesses of the particularistic approach can and will be eliminated provided the universalistic societies offer a better alternative worth emulating. For example, the Japanese are finally edging towards the removal of the particularistic inefficiencies of their service sector and in the process will move their economy to an even higher level.

To put it another way, the emotionalism of the particularistic approach is balanced by the pragmatism gained when situations rather than principles are the basis of action. Conversely, the scientific gains from pursuing correct principles in the universalistic approach are balanced by the dogmatic harm that results from undue attachment to inappropriate principles.

3
Japanese Industries To 1985

SHINZO KATADA

The Japanese economy in the first half of the 1980s will be characterised by two main features: a firmly rooted growth rate of 5-6 per cent together with a recovery in capital investment.

Despite a forecast of the same growth rate as in the past, the pattern of the economy is expected to change in the future. In the past seven years the economy was led largely by exogenous factors such as government capital investment and exports together with private capital investment growth of 1.2 per cent. In the future, however, private capital investment is predicted to grow at 7 per cent p.a. and government investment at 6.5 per cent p.a.; there will also be a continuing current account surplus.

In 1978/79, the Nomura Research Institute of Japan drew up a medium-term outlook for 80 selected Japanese industries. I have drawn from these in the following selected forecasts, each of which is based on the judgement of analysts of the basic indices for each industry.

There are five main factors influencing future growth:
Industrialisation of the service sector
Tendency to high value added
Changes in consumer needs
A switch from government investment to private capital investment
Completion of stock adjustment
Of these factors, the first three relate to structural changes and the others to cyclical factors.

Industrialisation of the service sector.   Consumer credit and leasing are typical industries benefiting from the industrialisation of services. As time passes and the number of listed companies increases, the restaurant industry, too, will come to the forefront. The following factors should be noted:

- The use of consumer credit as a means of planning purchases has begun to take root in Japan.

- The number of retailers who use consumer credit as a means of sales promotion is increasing.

- Companies granting credit are expected to have few problems in financing since the easing monetary situation with low interest rates is expected to continue.

In line with the growth of industry, newly listed companies such as Toshiba Credit, Central Finance and Daishinpan will appear in the coming three to four years following on existing listed companies such as Nippon Shinpan, Taces, Orient Finance, Hitachi Credit and Life.

Like consumer credit there is much room for the further development of leasing. The proportion of leasing in capital expenditure in Japan in 1977 was only 3 per cent compared with 12-15 per cent in the U.S.A. and over 5 per cent in the U.K., West Germany and France. The following factors will encourage the growth of the industry:

- It is expected that the Government will promote the spread of leasing in order to stimulate capital spending by introducing an investment tax credit in the early 1980s.

- The secondhand market for goods after the expiry of a lease has already been prepared so that transactions will be simple.

- There are many products in newly growing areas such as medical equipment and commercial and office equipment which are suitable for leasing.

- The financial base of leasing companies has been consolidated, following an increase in their profitability sufficient to encourage a diversification of their financial resources away from loans only to capital increases, bond issues and expected issues of long term commercial paper.

Tendency to high value added.   Machinery products with high value added such as medical equipment, computers, copying machines and automobile parts are expected to continue to grow rapidly.

Medical equipment

Factors contributing to growth are:

- The nation's medical expenditure is expected to continue to increase at an annual rate of 15 per cent due to the growing percentage of the aged in the population and the increased incidence of geriatric diseases and to the growing practice of undergoing early medical examination. The latter, together with the concomitant increase in the use of medical equipment, will to some extent take the place of drugs, whose share in total medical expenditure by 1985 is expected to have declined from the extremely high level of 37 per cent to the 25 per cent prevailing in the U.S.A. and European countries.

- The Government is promoting medical care systems such as local medical systems, the comprehensive medical examination system and the emergency medical information system, which will stimulate introduction of medical equipment on an extremely large scale.

- The introduction of medical equipment will enable hospitals to make savings on labour costs by systematisation of medical treatment and of administration.

- Producers of medical equipment are aiming at the improvement of the availability and quality of medical equipment through miniaturisation by means of the micro-computer and through simplification by introducing large sized machines or micro-computers. Examination equipment in particular is expected to show the most rapid growth and to have the largest market share in 1985, surpassing that of x-ray equipment, as a result of the much higher incidence of examinations than in the past.

Computers

Installed and operating computers in Japan are valued at 10 per cent of the total world value, next only to that of the U.S.A. So far the computer has mainly been used for centralised processing of data by a single, large machine. This concept is changing to that of decentralisation, or on-the-spot processing. It is expected that this change will encourage the growth of the computer industry.

Japanese-produced computers increased their market share from 51.5 per cent of total domestic demand in 1972 to 57.5 per cent in 1977. The top three companies in particular — Fujitsu, Hitachi and Nippon Electric — will continue to grow along with IBM, since: (1) their possession of semi-conductor plants enables them to reduce costs by means of semi-conductor technology; (2) there is much scope for the development of the export market; (3) under the guidance of MITI (Ministry of International Trade and Industry) the development of the new products for the coming period is progressing on schedule.

Copying machines

Exports are expected to be the leading factor in the growth of the copying machine industry. The world market is divided into two segments. Xerox has a commanding position in the market for high-speed machines for large consumers using more than 20,000 copies per machine per month, whilst Japanese products virtually monopolise the market for low-volume machines for small consumers. At present 70-80 per cent of newly installed PPC(Plain Paper Copying Machines) in the world are occupied by Japanese products. Copying machine producers such as Ricoh, Canon, Konishiroku Photo Industry and Sharp will continue to increase their shares in the PPC market by developing higher speed machines. In addition, these companies have diversified into areas such as facsimile (Ricoh) and computer-related products like LBP (Laser Beam Printer) (Canon) and the Ink Jet Printer (Konishiroku) which will further contribute to growth.

Automobile components

The replacement parts market is growing as a result of the increase in the stock of automobiles. The latter is expected to increase from 32.2 million units at the end of March 1978 to 42 million units at the end of March 1985. In addition, diversification of consumer demand and the tendency towards upgrading the requirements of cars will increase the rate of installation of optional components such as air conditioners, power steering and automatic transmission, whilst the strengthening of anti-pollution measures will encourage a rapid increase in components with electronic instruments. For example, the installation rate of Electronic Fuel Injection reached 30 per cent in Nissan's cars and 10 per cent in Toyota's cars in 1977, compared with 2 per cent and 3 per cent respectively in 1974.

Changes in needs.   The expected growth of these industries is largely a result of the industrialisation of the service sector or of the tendency towards high value added. In addition, however, growth will be stimulated by changes in needs. For example, the growth in consumer credit will mainly arise from a change in consumer attitudes towards it as a means of planned purchases. The growth of the automotive parts industry will also be partly connected to the diversification and upgrading of consumer needs.

Apart from these industries there are others which will be much more strongly affected by changes in consumer demand. These are the air transport, wrist watch, fashion and soft drinks industries, together with superstores and the processed meat industry, which have already been influenced by changes in consumer demand.

Air transport

As far as the business climate is concerned, of all Japanese industries, it is the air transport industry that will benefit most. An increase in passengers on domestic flights is expected as the interval between price increases on the Japan National Railways has and will become so short that the price of air transport will be cheaper in comparison. International flights will also benefit from the growing tendency towards leisure activities and from the price reduction resulting from the yen appreciation. However, the U.S.A. has been demanding the introduction of low air fares and the liberalisation of charter flights. Japan Air Lines whose dependency on international flights was 29 per cent in 1977 in terms of passengers, will be adversely affected by such liberalisation measures. All Nippon Airways, on the other hand, which specialises almost entirely in domestic flights, will benefit from the favourable business climate and from its own oligopolistic position.

Wrist watches

Demand for high quality and the emergence of electronics technology stimulated the appearance of a new product, the quartz watch, with one hundred times greater accuracy than that of the old mechanical watch and with the ability to indicate the time digitally. This new quartz watch firmly established Japanese prestige in the world wrist watch market. In 1977 Japan produced 44.7 million watches or 17 per cent of world production, exporting 28.5 million watches or 18 per cent of total world exports of 155 million. Although the domestic market reached maturity six or seven years ago, the quartz watch is expected to prompt replacement demand, since it can be supplied at a low price owing to mass production electronics technology. The annual growth rate is expected to be 10 per cent in both the domestic and the export markets.

Fashion

Until around 1975 the fashion market grew by more than 15 per cent annually, largely due to the increase in the fashion conscious population in their early 20s and to the immaturity of the ready-to-wear market. Since 1975, however, growth has slowed down as the population in their early 20s has decreased and the ready-to-wear market has rapidly reached the maturity stage, increasing its share in total ladies' suits sales from 41 per cent in 1970 to 71 per cent in 1975 and from 46 per cent to 63 per cent in sales of men's suits.

Instead, new, growing markets in sportswear, formal wear and clothes for married women will contribute to growth in the fashion industry. The sportswear market is expected to grow due to the fact that the variety of sports played by individuals has increased to include tennis in spring and autumn, swimming and climbing in the summer and so on. These games used to be confined to a limited number of people. The market for formal wear will grow as the tendency to throw parties at home becomes more widespread in Japan. As far as clothes for married women are concerned the market for large sizes is expected to grow. All in all, the fashion industry is expected to grow by around 10 per cent p.a. in the future.

Soft drinks

In the processed foods sector, soft drinks are expected to grow comparatively rapidly. The factors which will contribute to growth are as follows: an increase in winter demand due to the spread of central heating, an expansion of the domestic market through a strategy of making bottles larger, expansion of outdoor sales due to growing leisure activities and diversification of drinks from orange juice to apple, grape, tomato and vegetable juices, which will encourage new demand.

Superstores

In 1975 10 per cent of total consumer spending in retail outlets went to superstores, compared with only 3 per cent ten years ago. This tendency is expected to continue. The proportion should rise to 12 per cent in 1980 and 18 per cent in 1985. Although the introduction in 1974 of the Large Retailers Law affects the superstore sector by restricting the expansion of floor space in shops, it also encourages differences in growth and strategies between companies.

Those companies which have adapted to the Law through measures such as diversification or mergers and acquisitions will grow faster. Among them Ito-Yokado and Daiei have been diversifying into convenience stores and restaurants, whilst Jusco and Nichii aim to expand regionally through tie-ups or mergers with local superstores.

Processed meat

Westernisation of their way of life and the rise in their standard of living have induced the Japanese to increase their intake of animal protein, most notably of meat and processed meat. Whilst the consumption of rice, which was formerly the mainstay of the Japanese diet, declined from 257 kg per household in 1970 to 191 kg in 1977, consumption of meat and processed meat increased from 36 kg to 47 kg. This rapid rate of increase will continue until around 1982–83, when saturation levels of demand are likely to be reached.

Switch from public expenditure to private capital expenditure. In the medium term there are some industries which will benefit from the switch occurring in the 1980s from an economy led by public expenditure to one led by private capital expenditure. Among these will be electrical construction, air-conditioning installation work, general construction and road paving. These have so far benefited from public expenditure programmes and from electricity investment and will maintain their growth rates as a result of private capital expenditure.

Electrical construction

More than 50 per cent of the orders of electrical construction companies come from private electric power companies and Japan National Railways. Although the proportion of electrical construction in capital spending of the electricity industry will decrease as the latter's capital investment will tend to be centred on power generation sectors rather than on electricity distribution sectors, electrical construction for Japan National Railways and for the private sector will tend to increase.

Japan National Railways plan to build another five high-speed railway lines for the high-speed train (the 'Bullet') following the Tohoku and Shinetsu 'Bullet' lines which are under construction. In the near future private capital expenditure will contribute most to the growth of electrical construction companies since 30–40 per cent of today's orders of the large electrical construction companies arise from private capital investment, excluding that of electric power companies.

Air-conditioning installation

Until 1977 the private sector accounted for 70 per cent of total air-conditioning installations, in offices, hotels and factories. At the present time, however, the industry is benefiting from public expenditure and in the 80s will reap the benefits of increased private capital expenditure. In addition, this industry will be able to take advantage of new areas of equipment for which air-conditioning installations are indispensable, such as computer rooms, telecommunications equipment and nuclear plants, together with increasing numbers of air-conditioning systems run by local authorities and installations for subways.

General construction

As 60-65 per cent of orders come from the private sector, general construction is very sensitive to changes in private capital investment. So far, this industry has depended primarily on public expenditure programmes where profitability is somewhat inferior to that in private sector work, due to the local government policy of spreading orders evenly amongst local small construction companies. The private sector, on the other hand, usually prefers to give orders to large, general construction companies owing to their sub-contracting ability. In addition, there is much scope for general construction companies to expand in the profitable field of energy equipment such as LNG (Liquid Natural Gas) equipment, oil storage systems and nuclear power plants.

Road paving

Eighty per cent of this industry's contracts arise from public expenditure programmes and 20 per cent from private business in factories or plants. Although public expenditure programmes in the 80s are expected to grow less rapidly than in the late 70s, road investment, however, will grow by at least 10 per cent p.a. since the extent of properly paved roads in Japan is very low, being 40 per cent of total roads compared with almost 100 per cent in European countries. Particular emphasis will be laid on the improvement of secondary, local roads, since in 1976 only 29 per cent of these were paved.

Completion of stock adjustment.   Although the shipbuilding and electric furnace steel industries will grow rapidly in the future, these industries will benefit only from the upswing stage in the business cycle after completion of stock adjustment and a reduction in employees and capital equipment. Other structurally depressed industries such as petrochemicals, textiles and chemical fertilisers will, on the other hand, suffer almost chronic depression.

Shipbuilding

As a result of the construction boom in shipbuilding together with the subsequent slow-down in the world economy, considerable excess capacity in shipbuilding has arisen throughout the world. In 1977 30 per cent of world tankers were idle and it is estimated that future demand for new tankers will experience a dramatic fall from 23 per cent gross tons in 1975 to almost zero in 1980. Once the bottom is reached around 1980 and excess capacity is reduced, freight rates will probably recover in 1981 at the earliest, or around 1984 at the latest. However, since the difference in costs between Japan and Western European countries has been reduced and the shipbuilding industry in developing countries such as Korea, Poland and Brazil has caught up, recession in Japan's shipbuilding industry will continue.

Electric furnace steel

Amongst the structurally depressed industries, the electric furnace steel industry has adjusted most rapidly of all. Although excess capacity reached 35 per cent in 1975, it declined to a mere 10 per cent in 1978. This was a consequence of a decline in the number of firms from 58 in 1973 to 45 in 1978, of a 20 per cent decrease in the number of total employees and of a 20 per cent reduction in production capacity equivalent to 4 million tons.

The electric furnace industry will continue to fluctuate widely according to the phase of the business cycle since its commodities are highly sensitive to changes in the market. However, the industry will survive. Not only has it completed stock adjustment, but the cost of scrap, a main input of electric furnace products, is also tending to decline as a result of a substantial increase in the supply of scrap. Since the second half of the 1960s the accumulated stock of steel has increased by 30 million tons annually and massive amounts of scrap have begun to appear after the 13-year usage period.

★   ★   ★

Already since 1970, industries which underpinned the heavy industrialisation of post-war Japan such as iron and steel, petrochemicals, home electric appliances, heavy electric machinery and automobiles have been growing less rapidly than GNP. However, when looked at from the point of view of business opportunities, there appear to be some differences between process industries and basic manufacturing industries. Many basic manufacturing industries will suffer from low operating rates, loss of competitiveness and instability of profitability due to foreseeable fluctuations in raw materials costs. Process industries, on the other hand, have a number of opportunities in the form of promotion of exports, mainly capital goods, promotion of technological progress in the areas of home electronic appliances, telecommunications and electronic parts and upgrading and systematisation of automobiles, of sewing machines and of audio equipment. All of these developments have now been more or less completed.

The reasons why further development in electronics technology in Japan is expected are as follows: first, the necessity for information and telecommunications has been growing and secondly the accumulation of the technology of the integrated circuit (IC) in Japan is sufficient to ensure domestic production without recourse to imports. Until recently in the field of the IC Japan has been highly dependent on imports, mainly because Texas Instruments was the leader in the IC business owing to its strong price competitiveness. In fact in 1975 imports of IC's into Japan amounted to ¥40b. or 34% of domestic production of ¥117.6b. As the competitiveness of the Japanese-made IC has increased, this proportion has been substantially lowered, to the extent that in 1978 imports were only ¥50b. or 21 per cent of domestic production of ¥240b. whilst exports reached ¥44b.

Japan's competitiveness was strengthened largely owing to the fact that (1) technology which was accumulated in producing calculators has been applied to computer memories etc.; (2) the IC was designated as a key component of the computer industry by the Government under whose guidance Nippon Telegraph and Telephone Public Corporation and private companies have cooperated in developing modern technology; (3) since production of the IC requires massive capital spending and Japan was behind in setting up production, major electric companies were forced to make the IC themselves. Furthermore in developing and securing IC markets in competition with U.S.A. companies, Japanese companies were helped by the existence of their own distribution channels. As a result major electrical companies could invest the reserves earned from other divisions in their IC division and thus stimulate the introduction of automatic equipment and research and development, which greatly contributed to strengthening their competitiveness.

On the basis of IC technology Japan has a wide-ranging electronics parts industry, which particularly encourages technological advance in home electric appliances and telecommunications.

Home electric appliances

Just as the 1970s saw no leading products in the home electric
appliances market the 1980s too will lack leading products, apart
from Video Tape Recorders (VTR). However, the pace of
dissemination will be slow since, whilst the TV added vision as a
new medium of communication, the VTR lacks originality.
Consequently the rate of dissemination will not reach 10 per cent
until 1981-1982. Therefore in the medium term the home electric
appliances industry will aim to diversify its products and penetrate
further into other fields, such as industrial machinery or precision
machinery using its accumulated technology.

The diversification of products will mainly apply to durables such
as oil or gas heating, household goods, kitchen goods and so on.
Representative amongst these are, for example, room heaters using
petroleum as fuel (oil forced-ventilation system space heating
equipment), which have acquired rapid popularity, and electric
cookers. Being mass produced at low cost, these products are similar
to earlier, popular home electric appliances and manufacturers can
thus take advantage of their past accumulated resources by building
upon their brand image and using existing sales channels.

As far as diversification into industrial machinery and precision
machinery is concerned, companies are aiming to transform
products with mechanical technology into those with electronics
technology. By using electronics technology they were able to
produce office machines such as cash registers and calculators for
example. Recently they have been developing products such as
electronic wrist watches, electronic cameras and electronic organs.
The following fields are still under-developed and are expected to
expand as a result of utilisation of electronics technology: labour
saving machines such as numerically controlled machines or robots,
medical and health equipment, automobile telephones and communication equipment such as the home fascimile.

Telecommunications

As orders from Nippon Telephone & Telegram Public Corporation
(NTT) account for 45 per cent of total orders of the telecommunications industry, the expected slow-down in the growth of orders
from NTT will force telecommunications companies to develop
non-NTT orders. The fifth five-year programme to spread the use
of telephone equipment, which ended in 1977, achieved its target of
making the national telephone system fully automatic whilst the
number of telephones reached 34 million. The number of telephones
per capita has surpassed that of the U.K. and West Germany. As a

result, investment under the sixth five-year programme will be no higher than under the fifth, taking into account inflation. However, non-NTT fields — in particular exports and computer terminal equipment — are expected to grow rapidly. Exports of telecommunications equipment will continue to increase, mainly to the less developed countries, particularly Middle East and African countries. From the technical point of view, Japan is considered to lead the world in wireless equipment.

As for the domestic market, in addition to the growing field of computer terminal equipment, telecommunications companies can rely on others such as home facsimile, automobile telephones, marine communication satellites, on-line systematisation of the post office and so on. These new fields are expected to come to the forefront in the 1980s.

4

Japan And The New Industrial Countries Of East Asia

CHARLES SMITH

Since the first edition of this book appeared in late 1974 the world has come to accept that Japan's days of ultra-rapid economic growth are over. Japan will be lucky to achieve an annual growth rate of 5.7 per cent (the rate planned by the government) during the first half of the 1980s and may have to content itself with 5.0 percent or even less. This will be partly because energy shortages will be putting a tight rein on the economy until such time as Japan can reduce its dependence on the imported crude oil which now accounts for roughly 75 per cent of its total energy consumption.

Japan's shift to slower growth, however, has not been merely a response to energy problems. The Japanese economy today is mature in the sense that the material needs of the people have been satisfied. Concern with the quality of life has grown progressively stronger since the early 1970s and the Japanese are discovering, like the Europeans before them, that an improvement in quality tends to cost something in terms of quantity. Japan is apparently willing to pay that price: in other words its people have decided that more leisure time and a better environment are goals which take precedence over the mere maximisation of GNP.

Japan's attainment of economic maturity has coincided with the first flush of youth in a number of new Asian economies. The countries concerned include South Korea, Taiwan, Hong Kong and Singapore and could, in the next five to ten years, come to include other East Asian states, such as the Philippines, Malaysia and Thailand. The four most highly developed of the Asian NICs (New Industrial Countries) have enjoyed growth rates ranging between 6 and 12 per cent (real terms) over the past ten years and export growth rates ranging up to about 40 per cent (in the case of Taiwan and Korea). Per capita GNP in the four ranges from just under one quarter of the American level in Korea to nearly half in Singapore with the most advanced members of the group now enjoying income levels superior to those of Southern European countries such as Greece and Spain. The absolute sizes of the four economies remain small, but the point is approaching where the Asian NICs, together with the developing countries of ASEAN (the Association of East Asian Nations whose members are Singapore, Malaysia, Thailand, the Philippines and Indonesia) may exceed Japan's share of world trade within the next four to five years. They will do so thanks to their rapidly growing exports of manufactured goods many of which are now starting to challenge Japanese products in developed western markets.

The Asian NICs are already replacing Japan as suppliers to the West of black and white TV sets, toys, and textiles, while beginning to challenge the Japanese position in steel and shipbuilding. Within five years Korea and Taiwan are likely to have emerged as exporters of petrochemical products and of industrial plant — Korea in fact is already winning contracts for cement plants and sugar refineries. Once this stage has been reached Japan may have to start preparing itself for competition from the NICs in the motor industry and in advanced sectors of the electronics industry.

The NIC phenomenon has for some time been the cause of raised eyebrows in the developed West (including the U.K. where the Department of Industry published a report in January 1979 identifying no fewer than 23 NICs in Asia, Latin America and Eastern and Southern Europe).

What this chapter aims to do is not to discuss the phenomenon as a whole but to look at its significance for Japan and for the future pattern of economic relations between Japan and the West. The first aspect of the subject which deserves attention concerns the trade imbalance between Japan, the Asian NICs and the industrial West. Japan is in substantial surplus with all four NICs mentioned above and was still increasing its surplus rapidly with each of them up to the beginning of 1979. Meanwhile, no fewer than three out of the four (the exception is Singapore) have been increasing their surpluses with the U.S. and the European Economic Community.

From the point of view of the NICs the surpluses with the West and the deficits with Japan have cancelled each other out to produce a reasonable overall balance of trade. From the point of view of the West the deficits with the Far Eastern NICs have been an additional burden over and above that represented by bilateral deficits with Japan. Bilateral trade balances are not, of course, supposed to be matters of concern to free trading countries and the mere existence of imbalances with Japan and other Asian countries would not justify the amount of concern that has been expressed on the subject in the U.S. and Europe. What may, in part, explain western attitudes is the rate at which the deficits have been increasing and the rate at which exports from the NICs have been penetrating sensitive industrial sectors (as Japanese exports did a decade or so ago).

The explanation for the two sets of imbalances (between Japan and the NICs and between the NICs and the West) is relatively simple. Countries like Korea, Taiwan and Hong Kong focussed their initial sales efforts on Europe and the U.S. because they realised that the markets of these countries would be easier to penetrate than that of Japan, and because they were bigger. The same countries imported heavily from Japan from the start of their industrialisation programmes because Japan appeared to be a convenient and cost-effective source of machinery and industrial materials.

Geographical closeness gave Japan a big advantage in the competition to supply the Asian NICs but this alone might not have enabled it to achieve its present dominant position in the regional market. The *Sogo Shosha* (Japanese general trading companies) were quick to identify sales opportunities in neighbouring countries and to pass information back to their associates in manufacturing industry. At the same time both the *Shosha* and the manufacturing sector became large investors in the rest of the region, setting up plants which could only operate with capital equipment and materials supplied from Japan (either because they were obliged by their operating contracts to source their requirements in Japan or because the design of the plants and products made Japan an inevitable supplier).

Japan today is easily the largest foreign investor in South Korea and is well on the way to dominating the investment scene in Singapore. Japanese investment in Hong Kong is relatively small, but it remains large in Taiwan despite the political inhibitions caused by the normalisation of Japan-China diplomatic relations in 1972. Japan's investment performance in the region as a whole demonstrates the point that any advanced industrial nation wishing to trade actively with the NICs would do well to back up its trade promotion efforts by acquiring a manufacturing presence on the ground.

Having explained why the Asian NICs came to find themselves in deficit with Japan and in surplus with the West it is now necessary to point out how this situation could change, quite radically, in future. The Asian NICs are naturally aware of the dangers of exporting too successfully to the U.S. and Europe (some of these dangers, indeed, have already materialised in the shape of voluntary export restraint agreements on such products as textiles, shoes and colour TVsets). They are also (presumably) aware that Japan remains the largest consumer market in the world which has yet to be fully exploited by the exporters of other industrial nations. With this in mind the governments of all four countries have been placing increasing stress on sales promotion programmes aimed at Japan and their efforts, combined with those of private industry, are beginning to show results.

In terms of products, Asian NIC export successes to Japan include textiles (mainly from South Korea and Taiwan) watches (Korea, Hong Kong, Taiwan) furniture (Taiwan) toys (Hong Kong, Korea, Taiwan) and electronic products and components (Taiwan, Singapore, Hong Kong, Korea). Korea and Taiwan are also successful exporters of raw and processed foods to Japan and would no doubt be a good deal more successful if this were not a sector of the Japanese import market heavily protected by quotas and other kinds of non-tariff barriers.

In money terms, the evidence of the NICs export successes comes from Japan's own statistics. These show that, in the early 1970s, the U.S. was pre-eminent as a supplier of manufactured goods to Japan with Western Europe in second position and South-East Asia (including South Korea and Taiwan) a fairly distant third. Since the beginning of the decade the positions of Europe and South-East Asia have been reversed, with the latter holding 25 per cent of the Japanese manufactured goods import market (in early 1979) to Europe's 22 per cent. The U.S. still leads, but by a lesser margin than before, while a new residual group of countries, including China, has taken a further 25 per cent of the market.

During the first five months of 1979 when Japan's total import bill for manufactured goods was running approximately 50 per cent ahead of the levels of one year earlier, South-East Asia was recording gains of 60 to 70 per cent (varying according to the month) in the Japanese market while the export gains of Europe and the U.S. were in the region of 30 to 50 per cent. The South-East Asian exporters owed their successes partly to the price effects of yen revaluation (against their own dollar-linked domestic currencies) and partly to the declining competitive position of a number of Japanese domestic industries (also accentuated by yen revaluation).

The recent successes of NICs in penetrating the Japanese import market for manufactured goods raise questions for Japan and for its advanced industrial trading partners. For Japan the question is whether to allow the NICs free and unrestrained access to market sectors that were once the exclusive preserve of Japanese domestic industry. The answer offered by most government officials, economic commentators and representatives of large-scale private industry is that, within reason, the NICs should be given free access to the market. The well-known Japanese economist, Dr Saburo Okita, who has made a speciality of studying regional economic development problems says that the emergence of new industrial countries in East Asia should be accepted by Japan as an inevitable process and as a natural and reasonable one given that Japan itself was once a New Industrial Country.

Dr Okita and other Japanese opinion leaders, however, believe that the NICs can reasonably be asked to show some self-restraint in the manner or speed at which they allow their exports to penetrate sensitive sectors of 'mature' industrial countries' economies. Just how far such requests for restraint can reasonably be pushed, and what obligations are incurred by the mature industrial countries which make them are of course the crucial questions. Japan has no rule of thumb answer to either but it has shown what, by western standards, seems a reasonable ability to reconcile the conflicting demands of exporters in the NICs and its own sometimes vulnerable domestic industries in a way which holds out hope for dynamic change while rejecting the notion of fixed market shares.

The Japanese textile industry, which spearheaded the nation's industrial growth in its early phases, is a good example of adaptation to changing competitive conditions. The industry accounted for a dominant 18 per cent of Japanese exports as recently as 1967 but has seen its contribution shrink to a mere five per cent by the end of 1978 as the NICs took over traditional Japanese markets in Europe and North America. In Japan itself consumption of NIC-manufactured textiles now exceeds the value of textile exports out of the country despite the fact that in fashion goods and in mass-produced artificial fibres Japan remains exceedingly strong.

Shipbuilding is another industry which has been treated realistically in Japan. Its capacity has been cut down under a government-sponsored programme to one-third of the pre-1973 level while employment has been reduced by almost 50 per cent (from 87,000 to 50,000). This shrinkage has coincided with sharp increases in the capacity of the new shipbuilding industries of South Korea and Taiwan, although neither of these countries seems likely, for the time being, to challenge Japanese shipbuilders in Japan's own domestic market.

A third, more complex example of adjustment by a Japanese industry to the emergence of regional competition (or in this case potential regional collaboration) is to be found in electronics. No one supposes that the NICs are in any position to challenge the more advanced sectors of the Japanese electronics industry but some of its less sophisticated departments (monochrome TV, white goods, etc) have been almost literally 'exported' to the NICs over the past few years as big Japanese electronics companies have phased out their domestic operations in these sectors and established alternative facilities in neighbouring countries. Sometimes the export process has involved, not only finished products, but also a portion of the input into a finished product with the result that horizontal division of labour has begun to develop between the Japanese electronics industry and those of neighbouring countries.

One reason why the Japanese electronics industry has taken a positive attitude to the emergence of NICs is that it remains supremely confident of its basic strength. The big Japanese electronics manufacturers do not believe that the assistance they have given to neighbouring electronics industries will lead to an uncontrolled or uncontrollable invasion of the home market, for the very simple reason that newcomers to the market would be most unlikely to be able to break the Japanese industry's stranglehold on the domestic distribution system. The big Japanese electronic concerns, while tending to be coy about the reasons for European or American failure to penetrate Japan's electronics market, are almost brutally frank when the same question is asked about neighbouring Asian countries. Taiwanese or Korean companies can sell in Japan through OEM (original equipment manufacture) tie-ups with Japanese partners, they say, but stand little chance of penetrating the market under their own brand names (although some companies in both Taiwan and Korea are now trying to do just this).

A second reason for the confidence of the Japanese electronics makers is the extent of their technological lead. Most of the parts which go into Korean and Taiwanese colour TV sets (though not into monochrome sets) are still either exported from Japan or made by local affiliates or licensees of Japanese manufacturers. The technology tie-up with Singapore is even stronger, given the dominant Japanese investment presence in that island. In Hong Kong it has remained rather weak, but Hong Kong's electronics industry is not, at present, a competitor in the 'bread and butter' lines such as colour TV which are of prime importance to the Japanese electronics industry.

Japan's assumption of technological superiority vis-à-vis the NICs will only be reasonable, of course, for as long as Japanese industry continues to achieve advances as rapid or more than those being made in the NICs. This makes it imperative for the nation to generate new technology at a high rate throughout the 1980s. The accent is likely to be generation rather than acquisition now that Japan has virtually exhausted the reservoir of western industrial know-how on which it was able to draw during the first three decades after the end of World War II.

Sceptics about the future of the Japanese economy believe that an inability to produce original technology may turn out to be a fatal weakness in Japan's equipment for tackling the next phase of its development. Other, less sceptical, observers hold that Japan *will* succeed in generating its own new technology under the pressure of competition from its fast industrialising neighbours and that by doing so it will finally bridge what remains of the technological gap between itself and the industrialised West.

The final topic to which this chapter has to address itself concerns the impact on advanced western countries of Japan's rapidly strengthening relationship with the Asian NICs. From quite a number of viewpoints, this relationship has to be seen as fairly disturbing. Horizontal integration between Japan and the NICs in sectors like electronics will spell an increased ability on the part of East Asia to compete with Europe and North America in the sectors concerned. Competitive pressure on Japan to develop new technology so as to stay ahead of new industrial neighbours will spell still more competition (in areas such as computers) for the West. To cope with challenges such as these, the advanced western economies may need to adopt considerably more aggressive development policies for their advanced industries than they seem to be contemplating at present — or choose the alternative strategy of insulating themselves within a regional trading block (which in the long run might cost them dearly in terms of ability to compete in third markets).

The negative aspect of the challenge posed to the West by Japan's interaction with the NICs should not be underestimated, but there are positive aspects as well mostly related to the fact that the NICs themselves are unlikely to wish to become over-dependent on Japan. All four countries are aware already of the potential dangers involved in 'satellite status' vis-à-vis Japan (with the degree of awareness relating directly to geographical closeness). While this awareness persists businessmen from the West who arrive in developing Asia with offers of technology or investment are likely to find themselves welcomed with open arms.

The extent to which Western Europe has neglected the investment opportunities awaiting it in the East Asian NICs (paarticularly the two former Japanese colonial territories of Korea and Taiwan) is striking to anyone who has a nodding acquaintance with the region. British investment in South Korea at the end of 1978 was worth less than U.S. $10 million against a cumulative U.S. $350 million worth of Japanese investment. For Western Europe as a whole the number of technology licensing agreements in operation with Korean partners at the end of the same year was less than one hundred compared with Japan's score of 487.

Businessmen and governments in the region are perfectly well aware that much of the industrial know-how they have bought over the years from Japanese companies was acquired before that by Japan from Europe. There is a growing disposition in the region to by-pass Japan and deal directly with Europe especially where Japan is suspected of 'holding back' vital knowledge in order to protect its own competitive position. Electronic manufacturers in Korea and Hong Kong have tried to buy video-tape recorder technology from Japan but have been refused up to now on the grounds that they lacked the capability to make use of it. Alternative VTR technology could — and quite probably may — be made available to these countries by Europe.

The phenomenon of the emerging Asian NICs and their relationship with Japan can be summed up for the benefit of western business leaders in two simple propositions. The first is that nothing can stop it — not even the adoption by western countries of an out-and-out protectionist trading strategy. The second is that the situation is fraught with interesting opportunities, as well as disturbing challenges for mature industrial states. Japan has been quick to seize these opportunities, perhaps quicker than the NICs themselves might have wished. It is now up to Western Europe to claim its share of the action.

5
Working With Japan's Free Market Structure

GENE GREGORY

From since about 1976, the din of complaints from the United States and Europe about Japanese trading practices has grown steadily, and parallels the mounting Japanese balance of payments surpluses, the decline of the dollar and periodic upswings in unemployment that characterise the same period. The prevailing view is that all are linked, and the root problem is overly aggressive Japanese export promotions combined with protectionist measures which exclude foreign products from the burgeoning Japanese market. If the Japanese would only promote their exports less and open their doors wider to foreign imports, it is argued, all these difficulties would be more readily resolved.

On closer scrutiny, however, this perception proves illusory. If the view of Japanese trade practices is reasonably accurate for the 1950s and 1960s, when the economy was being rebuilt after almost total destruction during World War II, it does not reflect today's reality. Japanese trade and industrial policies put much less emphasis on export promotions and import restrictions, and devote much more attention to encouragement of imports along with adjustment assistance for Japanese industries no longer able to compete in markets at home or abroad.

Indeed, the central fact about the Japanese market is that imports of manufactures are growing both absolutely and as a share of total Japanese purchases abroad. Yet, remarkably, the United States, and to a lesser extent Western Europe, have lost some of their market share as suppliers of manufactured imports to Japan during the 1970s. In almost every category of manufactured goods, not just textiles, but also such sophisticated products as chemicals and machinery, market share has been lost to the industrial newcomers of East and South-East Asia.

The problem, then, is not that the Japanese market is closed to imports of foreign manufactured products; rather, it is that advanced industrial countries as a whole, and the United States especially, are losing market share due to declining competitive power. This basic pattern is not confined to Japan, of course. It is a worldwide phenomenon and the main causes are likely to be common to all markets. Alleged trade barriers in Japan most certainly do not provide a satisfactory explanation. It is the market mechanism, not its distortions, that challenges western enterprise in the Japanese market.

Common wisdom has it, for example, that the Japanese distribution system conspires, intentionally or not, to discriminate against foreign imported products. Apparent upon the most elementary examination, the fact that a number of U.S. and European firms hold major market positions in a wide variety of products would, in itself, seem to offer sufficient evidence that the distribution system in Japan, whatever its weaknesses, can be made to serve imports as well as it does local manufactures.

It is worth noting, after a more careful look, that this same distribution system has quite effectively been used to introduce South-East Asian manufactured products in direct competition with Japanese industries that are generally protected in most other advanced industrial countries.

'Calling the Japanese distribution system, which is a different way of doing business, a non-tariff barrier only confuses the issue,' Dr. H.F. Jung, president of Bayer Japan Ltd., says. 'The distribution system which has evolved in Japan over the years has its function in Japanese society. If a foreign businessman thinks he has a better distribution system he is free to prove that by setting it up and doing business successfully his way in Japan. Coca-Cola is a good example of that.'

Far from being a rigid structure which leaves the foreign firm no room for manoeuvre or innovation, as some critics claim, the Japanese distribution system is essentially dynamic, in a constant state of flux. To be sure, some sectors of the system are more flexible and dynamic than others. But, in general, the foreign exporter or investor finds that his options in the Japanese market are not dissimilar to those in other advanced industrial countries. The problem is to identify and understand those options, and this may take more time, effort and patience at the outset, simply because a new entrant into the Japanese market usually has less relevant experience than it does in other markets. Moreover, most foreign firms do not make the first and most important investment in language training for their managers, to equip them with the elementary tools for understanding and mastering the Japanese distribution system. Those who make this investment, however, are generally largely rewarded.

Olivetti is a case in point. Over the past 18 years, presidents of Olivetti Japan, equipped with a substantial working ability in Japanese, have developed a network designed to the company's special needs. Portable typewriters, once sold through the traditional office machinery wholesale channels, were switched to Olivetti's own direct-to-the-retailer sales organisation. Olivetti now has a comfortable 40 per cent of the portable typewriter market, which it gained against strong competition from Japanese manufacturers.

Direct sale to the retailer is also one of the main factors in the success of Estée Lauder's penetration of the men's cosmetics market. The company took charge of product distribution, from arrival in its Yokohama warehouses through to selling at its own retail counters strategically located in the men's department of major department stores. In so doing, it cut distribution costs far below those of its major competitors. As a result, Estée Lauder now boasts approximately 45-50 per cent of the total men's cosmetics business in department stores throughout Japan.

Taking a somewhat different route, Nihon Philips and Melitta both introduced coffee-makers into the Japanese market using coffee bean wholesalers and retailers rather than appliance channels that are largely controlled by Japanese makers. Practically speaking, the competition has been limited between the two European manufacturers. Although some Japanese appliance-makers have since introduced electric coffee-makers, neither retailers nor the consumers have so far taken them very seriously.

If such innovation in distribution is possible in Japan, however, this does not mean that existing channels are either closed or useless to foreign exporters. Nestlé has captured and held approximately 75 per cent of the instant coffee market (which accounts for 80 per cent of all consumer coffee sales) using the traditional wholesale network. Lipton's has become a household word in Japan synonymous with black tea, marketing its products through a joint venture with Mitsui and Co. and Toshoku, a major tea wholesaler.

Similarly, Johnson & Johnson has consolidated its position as a leading supplier of health-care products by working through the complicated existing distribution system. According to Kneale H. Ashwell, current president of Johnson & Johnson K.K., 'we sell none of our products directly. Instead, we sell through wholesalers, and sometimes through two or three different levels of wholesalers. In my opinion, Japanese wholesalers perform certain invaluable functions, such as pre-pricing, physical distribution, carrying credit risks, and so forth. What is needed is not to eliminate wholesalers, but to establish a more balanced relationship which allows us more interface with consumers.'

Under the guidance of its general agent, Chuo Bussan, Tampax, Inc. also sells its products through existing wholesale channels and is now the leading brand with a 30 per cent market share. Tampax has demonstrated, moreover, that the foreign firm is by no means at the mercy of established channels. To prevent the sale of its products at cut-rate prices, Tampax allows its wholesalers only a 5 per cent margin. But because of customer loyalty, assured through mass advertising and promotions, Tampax can exercise as much influence over its distribution channels as a major Japanese manufacturer.

Some foreign firms have found that the best channel into the market is through the distribution networks of Japanese manufacturers, who may or may not use existing wholesale channels. Pez, the Viennese maker of peppermints, packaged imaginatively to appeal to young people, successfully gained a position in the Japanese market through Morinaga, a leading confectionery company. Sony distributes Whirlpool appliances and Dutch-made Bruynzeel kitchen cabinets through its domestic trading company, while another Sony subsidiary, Sony Plaza Co., operates a chain of nine retail stores specialising in imported houseware, toiletries and sundries.

But Sony is not the only major Japanese exporter to turn its marketing energies to importing foreign products. Toyota's import arm, Toyoda Tsusho Kaisha Ltd., was founded as early as 1948, although it began to develop diversified trading in imported products only in the 1970s. With the same enthusiasm that has won it markets at home and abroad for motorcycles and cars, Honda has established four wholly-owned subsidiaris and one joint venture company in a serious bid for the growing Japanese market for foreign products.

Matsushita, which has gone about its import business methodically and rationally, offers distribution facilities for a vast array of foreign consumer durables, parts and components, raw materials, machines and measuring instruments. Two subsidiaries, Matsubo Electronic Instrument Sales Co. and Matsubo Electronic Components Sales Co., distribute these products, while two other subsidiaries market juke boxes, automatic ice-makers, washers and dryers, catering equipment and related products. Still another member of the Matsushita group sells products such as office-cleaning equipment to the service industry and consumer products to intermediaries for distribution.

The problem confronting foreign firms doing business in Japan is quite clearly not the lack of distribution channels; nor does the foreign marketer suffer any special competitive disadvantage inherent in the system itself. In general, the difficulties are of a much more subtle kind and derive directly from the foreign company's approach to the market.

Assuming that foreign companies have special products and services that are superior in quality or price, they will succeed in the long run only if they make a serious effort to market their products in Japan. If this seems an excessively obvious statement, it should be noted that many agents of foreign companies and their branch offices in Japan are wholly frustrated simply because the home office refuses to take basic differences in the Japanese market seriously.

Johnson & Johnson believe that it is very important to spend a lot of time and money in order to determine clearly what market segment to sell in. 'Product specialisation and concentration, rather than a broad, shotgun approach, are necessary in Japan.' And, since new product lift-off and payoff periods are longer than in the United States or Europe, 'patience, guts, and determination are required'.

As both Philips and Braun have learned, products often must be redesigned to Japanese specifications before they can be sold successfully. Both companies have reduced the size of their electric shavers to fit the smaller hands of Japanese users, and are now the market leaders.

Japanese consumers have an extraordinarily high sense of quality. 'As a result,' states Cornelis Bossers, president of Philips Industrial Development and Consultant Co. Ltd., 'the intense competition among manufacturers in Japan is such that only the best products will be able to maintain any kind of market share. Japanese consumers demand that products be flawless from the beginning. In Europe, for example, customers take for granted that about 3 per cent of all products will be slightly defective, and these will simply be serviced. In Japan, 97 per cent perfection is not good enough.' After-sales service is broader and more personal in Japan than in most countries: since such service tends to be costly, Japanese manufacturers find it cheaper to assure high quality standards to avoid after-sales service as much as possible.

Still, even after a foreign firm makes the necessary product adaptations and meets the high quality standards of Japanese consumers, there is a variety of problems that must be squarely faced.

Japanese firms tend to produce to very tight delivery schedules, for example. Also, since Japanese department stores do not maintain inventories, suppliers must have facilities near-by to ensure ready and steady supply, often making deliveries on a daily basis. Similarly, component and materials suppliers must maintain an inventory to assure the original equipment manufacturers or processors of regular deliveries to relieve the customer from the necessity of carrying the inventory himself.

Furthermore, if components and materials do not meet the users' tight specifications, they are returned, at the suppliers' expense and often with penalties imposed. Likewise, department stores usually reserve the right to return unsold merchandise and may impose penalties on suppliers for occupying their limited space with slow-moving items.

Obviously, not every foreign company is financially or technologically prepared to make or even interested in making the kind of effort these stringent practices require. The problem is not that the market discriminates against imported goods. On the contrary, in many instances there is a clear consumer preference for foreign products. But to meet that demand is often too expensive for the foreign exporter. There is, therefore, a strong temptation, to which many foreign firms have yielded, to try to enter Japan on the cheap. Products are assigned to a trading house or a distributor, prices are fixed to assure maximum short-term return, and the market is monitored by occasional visits from the home office.

This approach, however successful it may seem at the outset, is almost certain to end in disappointment. Either the low investment and high prices restrict the market for the product; or if the product has some advantage and can command a high price and high margins for both the exporter and the Japanese distributor, as the market widens it is certain to attract the interest of local competitors.

At this point, some critical decisions must be made. Either prices must be lowered sufficiently to discourage competitive entry, or the foreign firm must make a defensive investment in Japanese production to save and expand market share. Because the foreign exporter is much further from the market than the Japanese manufacturer, there is usually a lag in response to any crucial changes in Japan's competitive environment. The results can be disastrous. Foreign companies frequently introduce a product to the Japanese market and gain a substantial market share, sometimes from 80 to 100 per cent, only to see their position crumble suddenly before emergent Japanese competition.

The problem here is clearly not prejudice or discrimination against foreign imports. Rather it is the dynamics of a free market mechanism. Japanese manufacturers run the same kind of risk in this market, and successful Japanese firms must be constantly alert to protect their position through continual innovation and investment.

Before making the decision to invest in Japanese production, the foreign exporter should be aware of other problems he is likely to encounter due to differences in business practices.

U.S. and European firms tend to operate on a much shorter time cycle than Japanese companies. Since the foreign firm, and the American firm in particular, depends largely on capital markets for financing, managers are under heavy pressure to meet earnings per share targets, even on a quarterly basis. These pressures to provide increased earnings are great, and often mitigate against investments in the form of expense that must be covered out of operating profits. Thus, in general, the foreign firm tends to be concerned with short-term gain rather than long-term profits.

Japanese competitors tend to operate quite differently. Dependence on the capital market is less, to begin with, and in any case capital gains prospects weigh more heavily than short-term earnings per share in the assessment of corporate performance. As a result, although Japanese executives are closely monitored for their performance, their jobs do not depend on short-term indices. This explains, in part, why Japanese companies are prepared to operate on lower margins and tend to be concerned more with long-run than short-term profits. They are more inclined to make sacrifices of return on investment today for larger market share tomorrow. And this usually means that they are prepared to commit a larger share of their resources, at any given point in time, to investment in innovation and expansion.

If a foreign firm seeking short-term profit maximisation suddenly finds itself pitted against Japanese competitors working towards long-term market share, the results sooner or later are likely to be fatal for the foreigner. For the same reasons, joint ventures between foreign and Japanese companies often prove tenuous. If the foreign partner cannot make the adjustment to Japanese ways of finance and resource management, the joint venture is likely to lag behind the competition or there will be basic differences in corporate objectives between the two partners which will ultimately result in dissolution or withdrawal of one of the parties.

One solution to this problem, successful foreign companies have found, is to entrust management to competent Japanese executives and be guided by their experience and findings. They know the market place and how to compete in it. This does not mean, of course, that the foreign firm can dispense with assigning high level executives to supervise Japanese operations. Quite the contrary. Some remarkably successful firms, such as BSR and Melitta, base corporate vice-presidents or board members in Tokyo to direct or supervise their Japanese subsidiaries. An executive at this level of the corporate hierarchy, who communicates effectively, face-to-face with the Japanese management team, understands the demands of the market and can detect or anticipate changes in the business environment, is in a position to influence directly and expeditiously the decision-making at the home office to keep policy in tune with Japanese reality and in step with the rapid pace of change in the Japanese market. At the same time, this arrangement tends to speed the process of Japan's complicated approval procedures. Once again, these procedures were not conceived for the purpose of making life difficult for foreign firms operating in Japan, though they often have that effect because foreign business executives all too often do not make the effort to master the language and adjust to the Japanese business system. In the final analysis, this is the ultimate non-tariff barrier and it exists in the mind, rather than the eye, of the individual concerned.

The room available in the Japanese market for imports is, in fact, substantial and increasing. This means, of course, that Japan is likely to accelerate the process of replacing local manufacture of labour- and energy-intensive products with imports from South-East Asia, which suggests that some foreign firms might do better to consider supplying this market from bases strategically located in the region. But in general the trend towards increasing imports offers new opportunities to any foreign company prepared to make the necessary investment and adopt the long-range strategies demanded on the Japanese market.

6
Pointers To Success And Failure

SADAO OBA

Some success stories

To sell cosmetics to Japanese women is still a difficult business for foreign cosmetic manufacturers. Consequently, foreign-made cosmetics have got only 2 per cent of the total market, worth $56 million. Avon, however, has gradually established a foothold in the Japanese market and recently invested $11 million in a factory and warehouse as part of its first manufacturing venture in the country.

Marks & Spencer has made a sole distribution contract with Daiei, the largest supermarket chain in Japan. In August, 1977, as a trial order, Daiei bought a range of ladies', men's and children's clothes worth £500,000 and Daiei sold well by offering an alteration service for Japanese customers who are not as tall as Britons. This initial order developed into a sole distribution contract. Marks & Spencer produce the clothes according to the sizes supplied by Daiei and Daiei estimate the sale of Marks & Spencer's products at ¥500 million in the first year and ¥2 billion in the third year.

The sale of Britain's Tannoy hi-fidelity loud-speaker system to

Japan is something of a 'coals to Newcastle' story. In fact, Tannoy is exporting a sizeable part of its production to the most active hi-fi market in the world, Japan, where its executives spend a quarter of the year.

BASF Japan (West Germany's chemical giant's subsidiary) increased turnover by 9 per cent and profit by 60 per cent in 1978 by increasing the sale of imported chemicals and by the good results of their joint venture manufacturing company in Japan.

Licences and know-how are also in great demand. In 1977 Japan imported technology from West Germany, the U.K. and France worth a total of U.S. $165 million; this including technology for transportation equipment, chemicals, general machinery, electrical machinery, etc.

Why they have succeeded

Those successes were achieved not only by the strategies and efforts of the companies concerned, but also by the increasingly favourable environment for imported goods and the general operations of foreign companies in Japan.

Liberalisation in import and foreign capital investment into Japan has progressed considerably, especially since 1976 when the Keidanren Mission visited Europe where severe criticisms against the ever-increasing trade imbalance between Europe and Japan were raised. The trade imbalance issue was discussed also in 1977-78 between the U.S.A. and Japan and the Japanese Government was obliged to make various concessions in favour of both the European and American exporter. For instance, Japan's total imports (in dollar terms) more than doubled from $38,314 million in 1973 to $79,343 million in 1978. Imports from Western Europe increased in 1978 compared to the previous year. The increases are as follows: wool (2.5 times), pulp (7 times), aircraft (2.5 times), platinum (2.7 times), works of art and antiques (2 times), inorganic chemicals (83 per cent) and passenger cars (72 per cent).

Japan has skilfully managed to get rid of the stagflation after the first oil crisis of 1973-74 and, therefore, the yen has been continuously revalued against the U.S. dollar and European currencies. Because of this revaluation, Japanese consumers can buy imported goods cheaper than used to be the case.

The change in the attitude of Japanese consumers is significant. The traditional perception of imported goods — 'foreign-made products are good quality' — is still strong among the very sophisticated consumer products such as Scotch whisky, French perfume, Louis Viutton hand-bags, Burberry coats, or Borsalino hats. As regards ordinary consumer products, Japanese consumers pay less and less attention to their origin and more to their quality, design, price and the type of after-sales service being offered. One reason for this more sophisticated buying attitude of Japanese consumers is the growth in the number of Japanese tourists going abroad. At present, more than 3 million Japanese travel abroad every year with some 300,000 visiting Europe where they discover new products and a way of life which they have never seen in Japan.

Economic affluence, even in the age of uncertainty, changes the patterns of daily life in Japan. More and more women (including housewives) go out to work. The Japanese now eat out more often; there is an increasing demand for quick and inexpensive but highly nutritious meals. More and more the western diet of bread and meat replaces the traditional diet of rice and fish. To meet these demands, world-famous catering chains from Britain and America and even Czechoslovakian state organisations have established joint ventures with Japanese partners and are adding to the number of establishments in their chains.

Failures

But there is another side to the coin. Foreign products and companies have not always enjoyed such successes in Japan. Many prominent companies have, in fact, curtailed their activities or withdrawn completely from Japan.

British oil heaters had been very successfully marketed in Japan before local Japanese competition intensified. The rapidly increasing standard of living, however, created a need for more sophisticated types of heaters, and as a result British oil heaters lost most of their market share.

The export to Japan of Swiss watches (middle- to high-class) reached its peak of 821,000 in 1974, but has decreased every year since then to 578,000 in 1978. This is because the production of Swiss quartz watches was slow and could not cope with Japanese demand. The Swiss, however, have subsequently improved production of quartz watches and an export upturn to Japan of some 600,000 Swiss watches was expected in 1979.

Reasons for failure

●The usual complaint concerning the import of machinery from European countries is that European manufacturers do not keep delivery time as stipulated in the contract. This has often resulted in the loss of business.
●Neglect of the special or particular taste of Japanese consumers and

refusal to modify the product to meet the particular needs of the Japanese market.

- Neglect of the special characteristics of marketing in Japan, such as the role of trading companies and wholesalers, and the very complex structure of the distribution system.

- Disregard of the peculiarities of personnel management and labour relations in Japan. The life-time employment and seniority system is still powerful in the personnel administration of most Japanese companies; foreign companies may be required to do as the Japanese do.

- Inadequate knowledge of Japanese business practices. For example, the takeover, which is very common in the West, is rare in Japan; a takeover by force is generally regarded as an unfair and disgraceful practice.

- Underestimation of the strength of competitors. Japanese competitors are very quick to adapt their strategy in a new competitive environment once foreign competition comes in, and in many cases they have succeeded in producing in a few years, or even in several months, products more competitive in quality and price than the original foreign import.

- Refusal to make the effort to understand the Japanese mentality or to listen to Japanese advice. This can lead to the wrong choice either of the Japanese partner or of the foreign representative sent to Japan.

7
Japan's New Superconsumers

TERUYASU MURAKAMI

- The Japanese consumer market is continuously changing, and changing quickly and profoundly. The knowledge of a marketing manager who last visited Japan in 1971 is unquestionably inadequate for the establishment of a satisfactory marketing strategy to cope with the market today. Change has occurred in the market place, in the lifestyle of each group of consumers, in consumers' preference, consumer values, the level of disposable income, the pattern of usage of time and so on. Therefore, in order to be successful in the Japanese market, very close and constant market study is necessary. The fact that the size of the consumer market in Japan is far larger than the total oil revenues of all O.P.E.C. member countries put together will justify the expense.

- In line with the study of the market, the importance of advertising through the media, especially TV, should not be underestimated. In Japan, the amount of information on consumer goods is extremely high. In order to establish a well-defined brand and product image in Japan, an enormous advertising effort is required. If one looks at the successful top thirty foreign-based companies operating in Japan, one can see the evidence. These companies are all very keen on TV advertising and most of them are known even by seven-year-old children. A sweeping success, such as that of MacDonald's which opened the first hamburger shop in Tokyo's Ginza district in 1971 and seven years later had increased the number of its chain shops to over 150, could not have been obtained without intensive and efficient use of TV advertisements.

- Japanese consumers have grown to be highly sophisticated in discretionary items and highly selective in essential items. This phenomenon of dual polarisation can be particularly observed amongst the 'Bachelor Peerage', 'Bachelor Peeress' and 'New Family' groups (discussed in detail later), who are relatively high spenders. The case of Twining's Tea is the classic example of the success of discretionary goods. The company has succeeded in establishing a brand image of such quality that it makes people believe that only Queen Elizabeth and her family drink this distinguished tea. This corresponds particularly well with the image the Japanese have of the traditional New Year and mid-year gift they offer to their superior.

Some experts predict that Marks and Spencer will have significant success in the Japanese market if they can establish the same kind of image as they retain in Britain. St. Michael products are to be marketed by one of the largest supermarket multiples, Daiei. It is very likely that they will correspond well to the growing inclination towards high quality, competitively priced basic goods.

●Finally, it is very important to note that a rapidly increasing proportion of household expenditure is expenditure on services. Most consumer durables are reaching saturation level. For foreign exporters, higher graded products such as large-sized refrigerators still have a chance, serving replacement demand. If one wants to enjoy very high growth rates in the Japanese consumer market, and obtain the whole of the value-added element of the business, one must supply those goods related to the new type of services currently emerging, or carry out direct investment and develop such services from scratch.

Increasing size of the consumer market

The post-war rapid economic growth of Japan has been brought about by maintaining a dynamic balance between the extraordinarily high growth rate of private capital formation and the growth of exports that financed the foreign currency requirements for the importation of capital goods and technology. There is no doubt that capital formation and exports have been of far-reaching importance as major driving forces for economic growth. However, one should not ignore the importance of the post-war upsurge of private consumption expenditure that opened up a mass consumption market for mass produced goods, and thus promoted new and highly efficient capital equipment and technology. The expanded consumer market provided the foundation for the economy of scale of mass production systems, strengthening in turn the international competitiveness of Japanese products in the export market. The high economic growth gained through this process increased the income of Japanese consumers enormously.

In the mid-1960s the GNP of Japan was approximately only 10 per cent of that of the United States and far lower than that of West Germany and Britain. In 1977, the GNP of Japan was more than 40 per cent of the United States' GNP and greatly surpassed that of either West Germany or Britain. In line with this development, despite the traditionally high personal savings ratio of Japan, its consumption expenditure has grown greatly and now amounts to 446.5 billion dollars, second only to that of the United States.

Japan has the second largest consumer market in the free world. To give a rough idea of the size, it is 40 per cent larger than West Germany's and 2.8 times larger than Britain's, (based on private consumption only).

In terms of *per capita* consumption expenditure, Japan is ranked lower than the United States, West Germany, France and most of the Scandinavian countries. If one can assume that the level of *per capita* consumption indicates the level of maturity of the consumer, then the Japanese consumer is younger than his counterpart in those countries and there is still plenty of scope for further expansion of the Japanese consumer market.

Among the 32 million households in Japan the single person household accounts for approximately 10 per cent and the agricultural household for another 10 per cent. Of the remaining 80 per cent, 60 per cent are salaried and 20 per cent self-employed households. The average annual income of the salaried household is the lowest out of the three married households, amounting to 3.1 million yen in 1976.

However, the Japanese have tended to choose stability of life rather than the vicissitudes of self-employment and also urban life rather than rural. Consequently, the number of salaried households has increased more rapidly than that of other households. It is believed that at a certain point during the 1960s, the Japanese economy entered the Mass Consumption Society stage in Rostow's theory of the stages of economic growth. The cluster of households that pushed Japanese society from the Matured Society stage to the Mass Consumption Society stage and sustained it there was that of the expanding salaried household. This cluster has always been the target of the marketing manager of consumer goods manufacturing companies.

The consumption pattern of the salaried household is dominated by the summer and the winter bonus which together amount to more than 20 per cent of the annual income (in some cases it is as high as 40 per cent).

Changing patterns of household consumption expenditure

First, Japan is characterised by its high proportion of food expenditure, particularly in comparison with the United States and France.

Secondly, the share of clothing expenditure in Japan is slightly higher than in other countries whilst the share of housing expenditure is lower. Japanese consumers have also a greater propensity to spend more on clothing than do European consumers, including the West Germans who are noted for their high clothing expenditure.

Thirdly, miscellaneous expenditure is relatively low. Again, it is expecially so when compared with the United States and France, but it is quite close to that of the United Kingdom and West Germany.

Food is extremely expensive in Japan, especially meat and dairy products. Table 1 compares the retail prices of consumer goods in cities of four developed countries. The price of meat in Tokyo is almost three times that in New York, and potatoes are more than 5 times as expensive as those in Hamburg. However, relatively speaking, the cost of certain articles of clothing, colour TV and so forth is not as high as the cost of food. In certain specific areas, such as women's clothing and hair-dressing, Tokyo is cheaper than other places.

Another point worth remembering is the rapidly increasing expenditure on eating-out. Whilst the share of eating-out expenditure is only slightly above 10 per cent of total food expenditure, its growth rate in the seventies has been 17.3 per cent per annum, compared with 13.6 per cent for total food expenditure.

Miscellaneous expenditure is the only item which has increased its share in total expenditure, moving from 39.5 per cent in 1966, to 46.4 per cent in 1976. Since miscellaneous expenditure has both a high share in total expenditure and a high elasticity, its importance will increase in years to come. It is a very wide category and it is therefore difficult to foresee what will happen unless the contents are further broken down. Table 2 shows an international comparison of items of miscellaneous expenditure.

Private expenditure for health is higher than in the United Kingdom and lower than in the United States. This reflects the degree of development of the national health service in each country. In the United States, where most health care comes from private medical institutions, private expenditure is very high and amounts to more than 8 per cent of total monthly expenditure.

Expenditure on personal appearance and hygiene in Japan is very high and almost equals that of health expenditure.

The high figure for transportation and communication should be interpreted in conjunction with automobile-related expenditure. Throughout Japan the public-transport system is highly developed and it is far more convenient to use this system than to rely on cars. Automobile-related expenditure includes maintenance and petrol costs. The low figure for Japan can be explained by the low level of car ownership per household in comparison with that in the United States and the E.E.C., and by the lower figure for the average man-kilometre per person trip due to the scarcity of flat land.

Education in post-war Japan is something of a national obsession. It is believed that the only means of securing a decent future in Japanese society is to graduate from a national university with a high reputation. In order to do that, one must graduate with a certain standard from high school. Some secondary and elementary schools develop better reputations than others — even kindergartens do so! Long queues develop in front of kindergarten entrances when entrance examinations take place. Expenditure on education is one of the three items that have increased their share in miscellaneous expenditure over the past five years. It is 4.7 times greater than the figure for the United Kingdom. This cannot but be attributed to the differences in social systems.

The item 'others' accounts for various consumer durables. Even after the end of the boom period the highly acquisitive aspirations of Japanese consumers towards more and better consumer durable goods still maintain this figure at a very high level. It has decreased over the years as the penetration ratio of major consumer durables has reached saturation level. Table 3 indicates that manufacturers can only expect replacement demand for many of the listed goods. Some goods such as colour TVs and refrigerators have already reached an absolute saturation level. In the case of hobby or pastime related items, such as pianos and stereos, the penetration ratio will never reach the level attained by basic household goods.

Another problem for manufacturers is that the length of the replacement cycle of consumer durables is extending. The replacement cycle which started in the first quarter of 1970 took 7 quarters for one cycle. The following cycle which began in the third quarter of 1971 took 9 quarters. The next cycle which began in the third quarter of 1973 took 16 quarters. It is partly due to the changes induced by the oil crisis that consumers have become more savings-minded.

Apart from this rather temporary factor, there is a third problem for manufacturers. The latest trend observed since the beginning of the 1970s is that consumers are putting increasingly more emphasis on expenditure on services than on goods. The share of expenditure on services increased from 33.1 per cent in 1965 to 40.1 per cent in 1976. As a result of higher income levels, almost all households can afford day-to-day food and clothing, and they can secure a house in which to live and can obtain basic consumer durables. There is now a sound basis in money terms for the Japanese household to increase service expenditure. The incremental increase of income can be spent on discretionary consumption. The increase in free time of the housewife, mainly arising from the rationalisation of housework through electric consumer durables, together with the increasing number of companies adopting a 5-day week scheme, provide a sound basis in terms of time. Together these factors — money and time — encourage an increase in the service-content of household expenditure. Most service expenditure is classified under miscellaneous expenditure. Hence the most rapidly increasing item of household consumption expenditure will be miscellaneous expenditure in the years to come.

The consumer groups

The major historical factors which differentiate the generations are the Second World War and the rapid economic growth period of the 1960s. The most useful way of segmenting the consumer market is not by using urban/rural, rich/poor and graduate/non-graduate criteria but by dividing it into the generation before and that born after those two events. The difference between the consumption behaviour of the different generations is becoming more and more apparent. Consequently a marketing manager must always consider for which generation of consumers he is merchandising his product, and which is the most appropriate way of appealing to them in the light of their characteristics.

Adopting this approach, one can identify 7 groups.

Highly Matured Children
Bachelor Peerage
Bachelor Peeress
New Family
Middle Aged I
Middle Aged II
Old-Aged

Each cluster can be profiled as follows.

Highly Matured Children. It is rather early yet to identify this group as an individual segment of consumers, because they do not have a great deal of purchasing power of their own. The significance of this group is their capacity to exert a great deal of influence over their parents' decision-making when purchasing goods, especially consumer durables and hobby goods.

The word 'matured', does not refer to physical maturity, but to those children in late elementary school or early secondary school who have developed a mature knowledge of fashionable products such as sports cars, single-lens reflex cameras and video-tape recorders. Their knowledge of products does not end at being able to distinguish a Tannoy stereo speaker from a Sony one, but goes as far as distinguishing the technical differences between the two.

According to the *Guinness Book of Records*, the children with the highest IQ are Japanese. As a result of the Japanese education system which has stressed scientific education, a strong interest in the technical aspects of modern gadgets has developed. TV is another extremely important source of information. TV commercials on consumer products flood children's brains from 6 a.m. to 1.30 a.m. through five private broadcasting channels, reinforce their memories and keep them up-to-date.

Generally speaking, their parents in the Middle Aged categories I and II do not have up-to-date knowledge on new products and technology. This gap gives highly matured children a chance to intervene in the discussion between their parents on the purchase of consumer durables and hobby goods. The growing importance of this 'children's power' should not be ignored when knowledge and technology intensive products are introduced to the market.

Bachelor Peerage.   The inventor of this fascinating name, with its sarcastic undertone, must be a man in either the Middle Aged I or II category who feels somewhat jealous of the showy consumption patterns of the younger generation! The lucky members of the 'Bachelor Peerage' are the single, male population between 20 and 29 years of age.

The most notable characteristic of this group is its high disposable income. Due to the seniority system their annual income is not particularly high, at around £6,000 including tax. However, they usually live in their company's bachelor dormitory where the cost of accommodation and basic meals is extraordinarily low due to company subsidies. They do not have to worry about educational expenses for children in contrast to the Middle Aged I group, and expenditure for health can be managed within the national health insurance scheme. Hence, despite rather small nominal income, their disposable income is relatively high, often being higher than male disposable income in the Middle Aged I category.

Secondly, they are, first and foremost, the generation most deeply affected by the evolution of TV culture. The 'TV Child' who spent almost all his free time in front of the TV has grown up and become the 'Bachelor Peerage'. This generation places more value on feeling than thought. Feeling is the foundation of their life. Their sense of fashion is more elaborate than that of any other previous generation. They know what they want to consume, and the range of alternatives they can imagine as candidates for consumption is far wider than that of other groups. After all, they are the children of the so-called information society.

Thirdly, whilst they are concerned with job satisfaction, they are not at all 'workaholic' types. In the previous generation, the workaholic businessman was one of the ideal young men. A young businessman's duty then was to work hard, but a member of the 'Bachelor Peerage' is cooler in his ideas about working and can distance himself from work. In terms of consumer behaviour, 'Bachelor Peerage' employees have more discretionary time in which to consume more. These attributes of the 'Bachelor Peerage' group make them good quality consumers. Their purchasing power tends to contain a dual polarisation. The 'Bachelor Peerage' is very eager to buy expensive products and yet also very concerned to avoid expense on basics. Their motivation to buy new products is very high. This is the natural consequence of the second attribute of this group. There are hardly any psychological barriers when it comes to introducing new products into their lives. They tend to spend more on leisure and clothing. This is also a natural consequence. Because their disposable income is relatively high, they have a high sense of fashion, plenty of information about different lifestyles and enough free time in which to enjoy leisure.

Therefore, highly differentiated, high quality, expensive products which were believed to be only for professional use often find not a mass market, but a fairly sizeable market among the 'Bachelor Peerage' group.

Bachelor Peeress.   This is the single working-girl version of the 'Bachelor Peerage'. The basic characteristics of this group are similar to those of the 'Bachelor Peerage' group, though there are some differences.

Their age range is slightly narrower than that of the male version, probably being between 20 and 27 years old.

Despite the fact that their annual income is smaller than that of the 'Bachelor Peerage' group, the disposable income of Japanese single girl employees is greater, because they often live with their parents and thus do not even have expenditure on housing or basic meals.

Female employees have more free time than male workers. Leaving the office at the normal time, that is at 5 p.m. compared with the middle of the evening, has long been the established practice of girls in Japanese offices. Since working is always regarded as a temporary occupation which only lasts until marriage, Japanese girls do not worry about the company's view of their taking up all paid holidays or even taking more. Consequently many Japanese office girls take long holidays and travel in Europe, Hawaii and Guam. For the 'Bachelor Peerage' group, however, whilst they could afford to pay the cost of travel to Europe, they cannot afford the time for such a long holiday for fear of jeopardising their job or promotion prospects. There are no such restrictions on a 'Bachelor Peeress' and this applies not only to overseas travel but also other leisure and recreational activities. Learning Japanese flower arrangement, the tea ceremony and cooking used to be the most popular private adult lessons Japanese girls took before they married. As a result of the great increase in free time and money the private adult school market has expanded enormously. Now Japanese girls can even learn Spanish Flamenco dancing or field archery.

Fashion is, of course, their main concern. These girls are the fashion leaders, introducing new fashions such as the kitsch look, the layered look, the coordination look and the balloon shape, etc. into Japanese society. If a new fashion is not accepted by them, then there is little chance of it becoming popular throughout Japan.

New Family.   In the past as soon as a Japanese couple married, their life immediately tended to be the same as that of a married couple in their fifties. On marriage, the consumer's life was one of unchanged monotony and conservatism in clothing, eating habits and traditional role differentiation. Today, the typical 'New Family' scene on Sunday is that of a married couple, both in their early thirties wearing unisex sweaters and bleached-out jeans strolling hand in hand with their three-year-old child back to their flat in a large apartment block from the nearby supermarket. They are talking about the latest record of one of the punk rock singers, and their shopping contains a half-bottle of rosé wine produced in Japan, a baguette and some beef for a strogonoff.

Such couples were born during the 'baby boom' after the War. Their generation has always led the consumer market of post-war Japan. The educational market boom arrived when they were at school. When they began to get married, the marriage industry became very prosperous. When they started to build a house, the housing boom began. Now they have a family. Their family is a two generation family only, and they regard urban life as the only way of life for them. It is an extremely westernised or, to be more precise, Americanised way of life. The contemporary young couple in Japan places more importance on family life than on working life. The attitude is sometimes called 'my-home-ism'. The Japanese young couple also puts more emphasis on the satisfaction of spiritual needs than purely material needs.

As consumers, this generation's main concern is housing. More than in any other generation, their expenditure is directed towards furniture and furnishings. In contrast to the traditional Japanese social or business gatherings attended only by the head of the household, entertainment of families and close friends in the home is becoming increasingly popular amongst young couples.

In the past, once Japanese people were married, their consumption patterns immediately became very conservative. They considered new products to be for other people and ignored them. Today's 'New Family', however, retains its openmindedness toward new products and is quick to introduce them into the home. By the same token, the wife in the 'New Family' maintains her interest in the latest fashion trends. In their younger days, such wives were hypnotised by the mini-skirt revolution, when everbody wore a mini-skirt and it would have required a great deal of courage to wear a longer skirt in 1970 or 1971 in Tokyo. After a few years, however, everybody threw their mini-skirts away. The 'New Family' wife passed her college or office life during the dramatic rise and fall of this revolution and was an active participant in it. As a result of such close acquaintance with fashion and its influence on lifestyles, today's young Japanese wives continue to retain their interest in fashion and the fashion pages of women's magazines. It is said that one of the latest fashions in Japan, named 'New Traditional' or 'New Tra' was first accepted and developed by 'New Family' wives and not by 'Bachelor Peeresses'. If their influence continues to prevail, then the fashion market will have to examine this phenomenon much more closely and adapt accordingly.

Whilst they are keen consumers of new goods and services, 'New Family' wives also exhibit the dual polarisation consumption behaviour of the 'Bachelor Peerage'. Consequently increased consumption of basic goods cannot be expected.

The emergence of the 'New Family' is in itself of great significance for the Japanese consumer market. In addition to this, however, the influence of the 'New Family' is so wide-ranging that amongst the older generations, such as the 'Middle Aged I' or 'II', or even the 'Old Aged' generations, the emergence of the 'quasi-New Family' can be observed. Such people try to be younger, more fashionable and 'New'. Consequently a marketing strategy focussing on the 'New Family' could gain some enclave markets from amongst the other generations.

Middle Aged I.   This is the married, 35- to 44-year-old aged group, whose childhood and adolescence were spent in the war-time emergency economy and afterwards amidst the bombed ruins. These are the Japanese who put all their energies into the rapid development of the economy, changing Japan's ruined economy into the world's second largest super economy. Although the 'Middle Aged I' group earns £10,500 a year, their purchasing power is not at all strong, because they are often burdened with repayments of a house mortgage and notoriously high educational expenses for their children. The school enrolment fee itself is virtually nothing. What is expensive is the fringe parts of education. In some schools in urban areas, more than 90 per cent of pupils take private coaching lessons given by moonlighting school teachers or university students, to prepare for extremely competitive high-school and university entrance examinations. The fees for these private coaching lessons are so expensive that parents must forgo satisfying their own material demands. At present this section of society is one of the least interesting as consumers.

Middle Aged II.   This group also worked hard as instruments of the rapid economic growth. They may be entitled to call themselves the 'Lost Generation'. Their consumer life was totally sacrificed to fighting in the War and to reconstruction and development of the economy afterwards.

Now, however, they are enjoying the most stable and affluent period of their life. Their children are now economically independent. Thanks to the still persisting, though slightly weakened, seniority system, they earn enough to maintain a decent standard of living, that is as long as they are not affected by the creeping threat of a cutback in middle management as a rationalisation measure following the oil crisis. Their purchasing power is quite strong, although they are generally conservative when making decisions about purchasing new products. Their main concern is health. There is no hesitation in spending money on expensive medicines, health foods and health clubs. This health market is now a steadily growing one. A recently launched monthly magazine dealing with health care and promotion of good health gained an astonishing success. It is believed that the 'Middle Aged II' group are the main subscribers to this magazine.

Another of their concerns is old age. Because of the deficiencies of the national welfare system, they have had to prepare for old age themselves, but the means available to them have not brought them any great long-term security.

Old Aged.   The aged population of over 60 years has been negligible as a consumer. The aged used to live with their children, but the nuclearisation of the family in post-war Japan is changing their lifestyle in a gradual and profound fashion. A new mass of aged couples living by themselves is emerging as an independent cluster of Japanese consumers. So far, they have not been of any great significance. However, in the next ten or fifteen years the Japanese population structure will age at a speed three times that of European countries. As a result, the aged population will become a very important component of the Japanese consumer market. Whilst their disposable income is not high and their consumption behaviour rather conservative, yet the number of the aged population will increase so much that the Japanese consumer market will be forced to adapt. So far little is known about the future behaviour of this group of consumers but it will undoubtedly be one of the most important subjects of study for marketing managers of consumer products in the 1980s and 1990s in Japan.

The changing needs of consumers

The life of Japanese consumers has changed dramatically since the

end of the War. The consumer of 1945 could never have imagined
the affluence of contemporary Japanese consumers. Similarly, it is
difficult for younger consumers who only know the post-war chaos
by hearsay to imagine the life of consumers in 1945. The whole
process of change in consumers' lifestyle from 1945 to 1980 has been
an extremely complicated one. At the risk of oversimplification, one
can divide the whole period into three different stages. (1) from the
end of the War to the end of the 50s, (2) from 1960 to the oil crisis,
and (3) after the oil crisis, as shown in Table 4. Also one can
subdivide a consumer's life into six different categories. The needs of
consumers have gradually shifted from stage to stage.

As far as eating is concerned, in Stage I, and especially in the early
period, filling the stomach to its maximum was the consumer's
number one dream. As the reconstruction of the economy
proceeded, qualitative aspects began to be taken into account, but
only at the level of concern as to whether the food was calorie-rich or
not. A calorie-rich food was better of course than one low in
calories.

In the period of rapid economic growth in the 60s and of the
spread of westernisation of eating habits, Japanese consumers began
to eat more high-protein food, and to demand greater variety. As
more women went out to work, there was a growing tendency for
instant and snack food to be eaten. There were many advances in the
technology used in the food processing and frozen food industries
and in fish farming. Consequently many new instant foods and
snacks were invented, such as instant *ramen* — a bowl of instant
noodles — ready made curry, instant bean curd. As discretionary
income increased, eating-out expenditure grew. In 1965 eating-out
expenditure was 8 per cent of total food expenditure, and in 1970 it
went up to 12 per cent.

In Stage III, Japanese eating habits are becoming more and more
polarised. For day-to-day basic eating, the Japanese eat more
convenience foods than in the past, whilst on special occasions, they
eat much more in luxury restaurants and eat more high quality fancy
foods. Slimming is becoming an industry in Japan just as in other
developed countries. Increasing numbers of Japanese women,
middle-aged men and even children are paying more attention to
low-calorie food. Health-foods show the same trend.

After the upsurge of the movement against environmental
pollution in the early 70s, young educated Japanese housewives have
become very concerned about the safety of food. The consumer
movement greatly stimulated this concern. It became fashionable
for housewives to look into the contents of processed food in the
supermarket to see whether it contained preservatives and colouring. If so, she returned the product to the shelf with a frown. Such
scenes have been so widespread that food manufacturing companies

have been forced to lower the preservative and colouring content. Government regulations have become very stringent over chemicals in food. It is said that some confectioners' products in Europe cannot be exported to Japan because of the stringent control by the Ministry of Health over colouring. Whilst manufacturers may complain of this government control, regarding it as an invisible trade barrier, yet even if they could persuade the Japanese government otherwise, there would be little chance of convincing Japanese consumers.

In the area of clothing, securing neat and clean basic clothing for the four seasons in the year was the maximum demand in Stage I. In Stage II, consumers began to demand higher quality and greater variety. The traditional distinction between types of clothing according to the seasons gradually decreased in importance. In turn T.P.O. (Time, Place and Occasion) became the prevailing notions when purchasing clothing. Working/Leisure, Town/Home, Formal/Casual — the various categories of T.P.O. — created versatile sub-markets for clothing.

The introduction of the mini-skirt and the coloured shirt taught the Japanese the meaning of fashion. In Stage III, fashion is not the monopoly of the 'Bachelor Peeress'. Almost all clusters of consumers, from 'Highly Matured Children' to the 'Old Aged', are preoccupied with the question of whether the clothes they are buying are fashionable and trendy or not.

Recent closer links between Tokyo and other world fashion centres such as Faubourg Saint Honorè, Fifth Avenue, and South Moulton Street accelerate this tendency towards fashion orientation. Kenzo Takada and Issei Miyake's latest creations are shown without delay in women's magazines, such as *An-an, Nonno, Junon, More,* which are extraordinarily powerful media for trendy fashions and new lifestyles. Japanese women are famous in Paris for their craze for a certain brand of handbag. For some reason, this particular design of handbag suddenly became very popular amongst Japanese women, and Japanese visitors rushed into the Parisian shop selling them. The management panicked at this rush of buying by Japanese customers, and decided to impose an export control policy, selling only one handbag per Japanese visitor. It was rumoured that some Parisians began to hang around the shop offering to buy handbags on behalf of Japanese visitors for a fee.

In contrast to this orientation towards ready-to-wear fashion, an orientation towards more individual and more original fashion can also be observed. The layered look and the coordinated look have been followed by the 'Bachelor Peeress' and the 'New Family' groups. These fashion concepts are not to be found amongst ready-to-wear, off-the-peg clothes. The concepts merely recommend consumers to exercise greater individual and independent choice in clothing. This tendency is a growing one, as consumers become ever keener to create their own original fashion distinct from commercial fashion.

Housing is the latecomer to the list of changing consumer behaviour, because it requires much greater unit expenditure compared with other consumption items. But waves of change in consumer behaviour with respect to this item have also occurred.

In Stage I, a house meant space for basic needs such as sleeping and shelter. In Stage II, people regarded it as space for living. In Stage I, the plan of a house had been discussed in terms of the function of each room, such as the living room, kitchen etc., whereas in Stage II rooms were discussed in terms of the family members, such as the elder son's room, the youngest daughter's room. Privacy also became one of the important criteria for the Japanese consumer when looking for a house.

As far as housing is concerned, Japanese consumers can be divided into two clearly differentiated groups: those who manage to own their own house and those who do not. Rocketing land prices and construction costs have left many households as tenants of multi-story council flats or of rented houses. Japanese people have a very strong desire to own their own house, especially a detached house with its own garden. The home ownership ratio of the Japanese is relatively high but amongst the Japanese themselves there is strong dissatisfaction with the level of this ratio, with the number of rooms per house and the number of persons per room, which indicate the very low amenity level of Japanese housing. There is no doubt that housing is the least advanced area of the Japanese consumer's life. Those who do not own their own house and have given up the idea of ever owning one spend large amounts on other elements of housing, such as furniture and interior design.

In Stage III comfort has been given greater priority. As more households, especially the 'New Families', invite others to their home, the house is increasingly regarded as space for social life. Growing concern over security against burglary and earthquakes is also a new trend in Stage III.

Table 1   International comparison of retail price of consumer goods (1978)

(Unit:Yen)

	Unit	Tokyo	N.Y.	Hamburg	Paris
Beef	100g	308	108	159	171
Potato	1kg	175	72	32	41
Onion	1kg	91	170	128	144
Lettuce	1kg	242	201	346	198
Banana	1kg	180	113	136	243
Bread	1kg	289	270	296	108
Milk	200cc	43	16	20	19
Chocolate	50g	98	61	48	63
Instant coffee	150g	1,290	594	1,182	995
Codfish	1kg	1,220	944	978	1,081
Suit	1	45,300	41,000	37,000	48,900
Skirt	1	6,190	11,400	12,000	20,600
Colour TV	1 set	121,000	104,000	110,000	198,000
Refrigerator	1	104,000	126,000	106,000	51,000
Rent for private house	3.3m²/ month	4,230	6,030	1,880	4,600
Women's hairdressing	1	4,770	9,850	4,740	7,670

Table 2   International comparison of miscellaneous expenditure

	Japan		United States	West Germany	U.K.
	1972	1977	1972-73	1976	1973-75
Health	10.6	9.1	17.8	9.2	5.0
Personal care	10.5	9.4	—	—	4.2
Transportation and communication	} 12.7	} 13.9	4.8 5.5	4.5 4.8	7.7 3.6
Automobile related expenditure	10.3	13.0	48.1	34.8	35.1
Education	8.7	9.4	3.1	—	2.0
Smoking	3.3	2.5	4.1	—	11.4
Reading	6.7	6.1	1.5	19.8	—
Travel	5.5	5.5	3.0	3.1	—
Others	31.1	31.0	12.1	23.8	30.9

Table 3 Household penetration ratio of consumer durable goods

(Unit:%)

	1965	1970	1975	1977
Colour TV	—	26.3	90.3	95.4
Stereo	34.6	31.2	52.1	54.9
Radio	74.0;	71.7	77.4	77.8
Camera	69.3	64.1	77.4	72.1
Washing machine	72.7	91.4	97.6	97.8
Hoover	41.4	68.3	91.2	94.4
Refrigerator	62.4	89.1	96.7	98.4
Electric fan	70.7	83.2	94.3	94.9
Air conditioner	—	—	17.2	25.7
Paraffin stove	37.7	79.1	89.0	89.8
Home organ	11.0	17.4	22.9	23.9
Piano	3.2	6.8	11.8	13.0
Bed (western style)	—	23.9	37.8	44.7
Car	—	22.1	41.2	44.9
Motorcycle and motorscooter	28.1	27.2	21.0	21.3

Table 4   Changes in consumer demand

	Stage I 40s and 50s	Stage II 60s and early 70s	Stage III Late 70s and 80s
Eating	Quantity High calorie content	Variety High protein Snacks & instant foods Eating out	Eating as leisure Lower calorie content Safer food, health food
Clothing	Bodily protection Keeping warm Basic clothing	High quality Variety T.P.O.	Fashion Individuality Internationalisation
Housing	Space for sleeping Shelter and basic utilities	Space for living Privacy	Space for social life Comfortable house Security
Health	Passive healthcare Cure of disease	Active healthcare Maintaining health Preventive care	Healthcare as leisure Building-up of good health
Leisure	Resting Passive leisure Indoor activity	Participative leisure Expensive leisure Outdoor activity	Diversified leisure Individuality Social life
Education	Compulsory education Vocational education	Higher education	Life-long education Education as leisure

8
Rationalising The Distribution System

MASAO OKAMOTO

It has been said that the complexity of the Japanese distribution system has led many foreign companies to forego or fail in their attempts to penetrate the Japanese market. At its worst, the 'inexplicable' layers of secondary and tertiary wholesalers appear to be a conspiracy to exclude imports. It seemingly buttresses the image of 'Japan Inc.' — the myth that all industry and government are constantly scratching each others' back in one cosy harmony and unison.

Yet, few seem to realise that the Japanese manufacturers also face the same obstacle of a complex distribution system. After achieving high efficiency in production procedures, the next target of the manufacturers should be the rationalisation of a distribution system to obtain further productivity or profitability. So, in the future, we can expect to see Japanese manufacturers distributing their products far more effectively as well as developing a more efficient system for collecting and exploiting marketing information through rationalised channels.

The movement towards rationalisation of the distribution system is also found on the consumer side. Japanese consumers claim that higher commodity prices are due to the complex distribution system and that immediate rationalisation is vital. The problem of distribution in Japan has been the bane of Japanese manufacturers and consumers alike for a long time; as an obstacle it is clearly not confined to those outside looking in.

So what can be done to rectify this situation? How can the Japanese private sector help foreign companies advance into this market of 115 million? If more foreign enterprises used their imagination more and made a sustained and determined effort, the private sector in Japan could be of considerable help in overcoming the distribution dilemma. Before developing this idea further, let us first explain briefly how the current distribution structure evolved.

During the 1950s, the retail sector was highly fragmented and inefficient. The vast majority of retail outlets were marginal units managed by 'Mom and Pop' teams. The high growth in the 1960s seemed set to implement a thorough overhaul on the assumption that the relatively slow rates of productivity increase in the retail sector, combined with the upward pressure on wages, would crowd out the marginal units and lead to rationalisation. But, in the event, this phenomenon, familiar in the West, just did not happen in Japan.

In order to match increasing mass production with mass consumer sales, the manufacturers pumped the retailers with rebates, discounts and easy finance terms; also the rising wage rate obviously did not have much effect on the 'Mom and Pop' operations. In fact, instead of being crowded out, these establishments flourished with the rising consumption expenditure; for example, between 1960 and 1969 when G.N.P. increased by 3.59 times, retail sales grew 3.57 times.

During the 1970s, the drastic cut in consumption expenditure following the oil crisis and the increasing displeasure of the manufacturers with the terms of business should together have cut into the marginal outlets. Again, however, that just did not happen. Due to the shopping habits of the average Japanese household, despite the increasing ownership of cars, the best retail areas continue to be near commuting railway stations and bus stops, and that is exactly where the small proprietors have been established for years.

In competing for this limited space, a volume sales store must set up operations by purchasing or leasing land which is currently valued at ten or more times the value compared to the prices paid when the small proprietors entered the area. Due to this enormous differential, which directly affects the tax rates, the large volume outlets have yet to solve completely the problems of achieving higher returns than the smaller proprietors.

Buying out the small proprietors is not as simple as it may appear: there are the added problems of finding alternative accommodation since most of them live on the second floor of their stores; what is more, their average age invariably precludes them from changing to other forms of livelihood. Increasingly, the small retailers do not consider their returns on investment in deciding whether or not to remain in business.

Rather, they have been using their considerable political weight to curb the advance of the volume sales stores. During the fifties, the most salient threat to the small outlets came from the department stores. Hence in 1956, they successfully lobbied for and won the *Department Store Law* despite criticism from big business calling it 'a commercial policy designed as a social relief measure'. And three years later, the lobby repeated its success with the *Retail Control Special Measure Law*.

By the end of the sixties, there appeared a new threat — this time from the supermarket chains. Here again, the small retailers were successful in enacting *The Large Retail Store Law of 1974*. The law requires chain store operators with more than 1,500 square metres of floor space to negotiate with merchants already in the area about such matters as the number of days the store can be open, business hours, and the store size. The government would like to repeal some of these laws but the chances of doing so are not very great.

From this standpoint, the Japanese distribution system offers a close parallel with its agricultural industry. They are both mostly managed by 'Mom and Pop' teams, plagued with marginal small units, but nevertheless entrenched in their positions through the activities of their powerful lobbies and the promising price of land on which their establishments stand. Smaller neighbourhood outlets are showing tremendous tenacity and are not yielding as rapidly to the advances of the volume sales stores as in some western countries.

Statistics bear out this dominant profile of too many, too small, too specialised and highly marginal retailers. There are approximately 1.4 million retailers in Japan. According to recent figures, the number of retailers per 1,000 people is 18.89. This is almost twice the rate for the United States, Britain, and Finland, which were respectively, 9.08, 10.9 and 8.09. Furthermore, over 60 per cent of retailers have only one or two employees, meaning mostly 'Mom and Pop' operations. If we include those with three to four employees, the percentage is over 85 per cent. From the other end of the spectrum, those with over thirty employees still hold less than one per cent of the total number of retailers.

There are no comparable figures on specialisation, but government publications suggest that the incidence of specialisation is higher in Japan than in other countries. For example, butchers in Japan have about 80 per cent of their sales in meat items. Tea wholesalers sometimes have an even higher percentage of their sales in tea alone. This high degree of specialisation is made tenable by the fact that stores of various types tend to cluster along the main commuter rail and bus routes.

With respect to sales per employee, available figures are quite revealing. The average annual sales for employees of stores with one or two employees was around 5.2 million yen in 1976. For employees of stores with more than 100 employees, it exceeded 21.3 million, or more than 4 times the sales of the former group. However, only one tenth of one per cent of retail establishments had over 100 employees while, as mentioned previously, the former group held over sixty per cent of the retail sector.

These characteristics have made wholesalers extremely important in Japan, and explain some of their functions: the margin of risk that individual retailers can take is very small, their financial resources are limited, and they are extremely cautious about allocating their small display and stock room space. Consequently, wholesalers usually agree to share some of the retailers' risks and essentially act as hand-holding partners.

A good example here is the widespread practice of selling goods to retailers on consignment. If the retailer finds that the goods are not moving, he reserves the right to return them to the wholesaler. Sometimes, the wholesaler can, in turn, return them to the manufacturer, although he may accept some of the manufacturer's risk.

In addition, the wholesaler will often provide credit to the retailer. Goods are normally sold on notes for periods from 30 to 90 days. The wholesaler generally provides some, or in many cases, all the financial resources to finance this line of credit.

Furthermore, when new products are being introduced, it is often the wholesaler who tries to convince the retailer to allocate floor space. Their constant and frequent calls, necessitated by the usually small stock capacity of the retailers, obviously make the wholesalers a very useful agent for the manufacturers.

All this requires constant contact among the retailers and the wholesalers. The wholesaler must always maintain the best of relations and vice versa. When the number of retailers is large, this calls for a larger number of wholesalers than one may expect in other countries. And when we consider the fact that many of the wholesalers are in the same boat as the retailers in terms of their financial resources and the size of business, we must also take into account the wholesalers' wholesaler, and so on down the line into this complex distribution web.

One generalisation which too often holds true in Japan is that the smaller the producer or the retailer, the more the number of intermediaries there are between the producer and end user. In many cases, paying the wholesalers' margin is the only feasible alternative, unless the new entrant is willing to make a major investment in setting up his own separate distribution channels.

Of the estimated quarter million wholesalers in Japan, a full one-fifth have only one or two employees. Those with nine or less employees still account for more than seventy per cent while those with over 50 employees hold only about three per cent of the wholesale sector.

In contrast to the retail sector, however, there have been some encouraging signs of consolidation in the wholesale sector. There are several reasons here, and several promising areas where foreign companies can make their contributions.

First, compared to the 1960s, many more manufacturers are becoming strong enough to establish their own distribution systems. Apart from creating captive sales companies, they have been encouraging mergers among wholesalers and many of them have been responding by establishing wholesale chains. Second, the large trading companies have begun to infiltrate the wholesale area and have been rationalising their operations. Third, despite their numerous obstacles, the growth of the volume sales stores has, in turn, encouraged the concentration of wholesalers. Finally, the effects have been magnified by the fact that larger and rapidly expanding wholesale operations have been experiencing more rapid increase in productivity.

Although the fragmentation of the retailers will continue to necessitate hand-holding wholesalers, there is still room for more rationalisation of the wholesale sector. Indeed, it is one objective the Japanese private sector must continue to strive for in order to achieve a more efficient economy. And it is also one of the major developments that will assist foreign companies trying to sell in Japan.

The question then becomes: how should foreign companies make their entrance? How can they participate in these developments and benefit from them? There are several avenues open to foreign companies and there is a wide variety of choice. A check list of possible avenues is as follows:

- Large general trading companies.
- Medium to small trading companies.
- Specialty trading companies.
- Large retail outlets like supermarkets, department stores and self-service chains.
- Non-Japanese trading companies.
- Private commission agents.
- In-house or captive trading and distribution companies.
- Japanese manufacturers of complementary line products.

Schemes contemplated under this last item would be, for example, selling American refrigerators through the distribution network of Japanese T.V. manufacturers. Another possibility would be to link up with Japanese plant export contracts by providing parts for the plants. Yet another possibility may be to have the Japanese manufacturer incorporate foreign-made components into its final products like European spark plugs for Japanese automobiles.

The experience of various multi-nationals suggests that the selection of the avenue that best fits the product is the most important part of selling in Japan. For example, should you depend upon large trading companies or specialised trading companies? Should you rely upon your own captive distribution company or the

large retail outlets? Sometimes even the selection of the right avenue
in the short run may have adverse effects in the long run. Deciding
on the right avenue is a vital requirement when selling in Japan.

Let me illustrate these points with some company success stories.
The experience of the Philips Corporation suggests that goods that
require extensive after-sales service to customers should be handled
by an organisation that can be more responsive to the needs of each
merchandise than the giant trading companies. Although in the
short run Philips was able to get their foot in the door, they
ultimately decided to establish their own distribution system. On
the other hand, fast food products such as Kentucky Fried Chicken
have been enjoying enormous success after tying up with the
Mitsubishi Corporation.

Another alternative to establishing your own distribution network is to use the network of a Japanese manufacturer of similar
goods. For example, while G. F. had only a 5-10 per cent share of the
market with their own distribution system, they were able to
increase their share to 20-25 per cent after tying up with the Japanese
seasoning manufacturer, Ajinomoto. An even more spectacular
expansion of market share was achieved by Warner-Lambert with
Hattori Tokeiten. Through Hattori's distribution system, Warner
Lambert was able to increase its share of the $40 million safety razor
and blade market to 65 per cent having outstripped both Gillette and
the once dominant domestic Feather brand.

The common denominator underlining all these success stories is
the fact that all the foreign firms had made a determined effort to
understand the Japanese market, decided to dig in for the long haul,
and waited patiently for their returns. The problem here is that such
strategy goes against the grain of foreign management style.

With their home ground the largest market in the world, many
American producers have failed to learn the requirements of export
management. Too often they have taken a 'take it or leave it' attitude
towards merchandise designs. Export models must be designed as
export models. Even European producers, used to international
trade, should cultivate new export tactics for the Japanese market,
because the Japanese market has different characteristics from those
of Europe, Africa and the Middle East. Behind Philips' success there
was extensive market research to determine the proper size for their
home electric appliances which must as a general rule be smaller than
in western markets. Even Coca-Cola changed its product ingredients
and taste to sell in the Japanese market.

Furthermore, foreign companies are often too eager to show early
returns to satisfy their investors. One reason why Japanese
companies have been successful in exporting is that they are not
restricted by such considerations. Export management requires long
term commitments before returns start coming in and the balance

sheet must be watched accordingly.

There are limits to what the Japanese private sector can accomplish. There are social, political and historical factors behind the inefficiency of the distribution system. Although its political weight precludes an early solution to this problem, there are market forces at work to rectify the situation, and foreign companies can hope to benefit from these developments. In the final analysis, however, the burden of initiative on foreign management cannot be denied nor neglected.

9
The Value Of Market Research

ANDREW WATT

If you want to find out about consumer behaviour and attitudes in your company's home market, the chances are that you will hire a professional market research agency to ask the questions. So what do you do in Japan? Ask the taxi-driver on the long journey from Narita airport to Tokyo? Check things out with a secretary in your Japanese distributor's office? Take as gospel a casual comment from a Japanese wholesaler? Of course not. But the sad fact is that many European executives do willingly 'suspend disbelief' as soon as they get off the plane in Japan. They have been so brainwashed about the difficulty of understanding the 'inscrutable' Japanese market that they place themselves at the mercy of anyone who seems to speak convincingly about 'the Japanese', instead of trying to find out the facts for themselves.

Yet objective market research is needed far more in a culture that is markedly 'different', like Japan, than in one's own domestic market. In fact, Japanese domestic firms themselves do market research among Japanese consumers — why should foreign firms handicap themselves by depending on anything less objective? And indeed, as one non-Japanese executive remarked, market research data can be the foreigner's secret weapon in Japan. By finding out the facts, and examining the Japanese consumer without pre-judging the situation, and without hang-ups, he can balance his own lack of cultural background, and obtain a surprisingly clear picture of what makes the consumer tick.

The tools are available

Market research in Japan may not be as well developed as in Europe or America, at least in terms of volume or the number of people and firms involved. (The Japan Market Research Association, which includes almost all major firms, numbers only thirty-nine members.) But it is no struggling infant. The companies involved have been operating for some time — our own organisation, for example, opened its research operation in 1957, and other firms were ahead of us. There is a steady flow of technical information from other countries, so that Japanese researchers are well in touch with the latest developments elsewhere. And they have adjusted techniques to the purely local aspects of the research scene. It is worth mentioning that there are several international research agencies with full-scale operations in Japan, which can provide a bi-cultural awareness, on top of regular research expertise, that can ensure the best value for money.

Summarising the Japanese situation in practical terms, it is worth first pointing out the plusses: respondents are cooperative, well-educated and literate; social taboos do not prevent interviewing; the streets are safe, so female interviewers can conduct evening interviews in any city area; everyone speaks the same language, and regional dialects present few problems; interviewers can be found who are themselves intelligent and conscientious.

But there are difficulties that need to be kept in mind.

Courtesy bias can lead to a respondent giving the answer he or she feels the interviewer wants to hear.   We have plenty of cases in our files which underline this point.

People feel uneasy in the presence of those they do not know, and may not state their opinions in a forthright way.   In a face-to-face interview, the fact that the interviewer is a stranger is itself reassuring to the respondent, and even quite personal questions will be answered. In a group discussion, the warming-up part of the session may take longer than in Europe, but, given a skilful moderator, contrasting opinions do emerge.

The language is indirect and imprecise — and hard to translate.   This is certainly true. Japanese is a language full of unfinished sentences whose full meaning the listener has to fill in for himself, where a negative feeling may be expressed by a neutral comment, and where a literal translation may often give a misleading impression of what is being said. Professional help is available to explain in English the real meaning of what is said in Japanese. This cannot be totally distortion-free, but the international research agencies have lived with this problem for many years, and have learnt to bridge the cultural gap.

Social norms as perceived by the respondent will affect his or her answers. The Japanese often talk about the difference between the outside and the inside — between the formalised, generally accepted response (*tatemae*), and the real feeling inside (*honne*) — and this certainly affects research interviewing. But it is worth noting that this is not an exclusively Japanese phenomenon. In addition, these perceived social norms also influence, consciously or unconsciously, the way people behave, the products they buy, and so on. They do not just affect research; they are relevant in other contexts too.

Practical differences

Anyone involved in research in different countries has probably already given up the effort to standardise methodology from market to market. (What is important, in the words of John Downham, is to insist on high *standards* everywhere, not *standardisation*.) In Japan fieldwork methods are determined by cultural patterns and other facts of life that cannot be changed.

Location. In Japan, there is a marked territorial difference between 'inside' and 'outside' the home. The border is not the actual doorway; the 'outside' area continues inside the front door, and then there is a step up, or some other line of demarcation, to indicate that beyond that point is the 'inside' of the home. When you 'go up' (*agaru*) into the inside area, you take off your shoes, and then you are really inside. But a research interviewer would not normally get that far. The interview is conducted either standing at the front door, or more usually just inside the front door, in the entrance (*genkan*) area that still counts as 'outside', and with the interviewer keeping his or her shoes on. The respondent will probably sit in a kneeling position just inside the demarcation line. All this means that interviews are not conducted in the most comfortable of environments. And for an interviewer to ask to 'go up' would be a breach of etiquette, while the respondent is unlikely to make such an invitation. Exceptions to this Japanese version of the home-castle syndrome are very rare.

Parking. Japan's cities have one of the world's worst parking problems — largely made up as they are of narrow streets or lanes

where there is barely room to drive, let alone park a car. Because of
this, although many interviewers may have a car, it just is not
practical to use it. Interviewers, therefore, have to use public
transport which is fast and convenient, but of course does not take
one from door to door as a car would. So if a client requires
interviewers to carry something heavy or bulky, the reception will
be far from enthusiastic. In some cases, we have a male student or
part-time worker to help carry whatever it is; —but of course that
increases fieldwork costs.

A gift.   A gift is mandatory at the end of each interview. A
handkerchief set, a pen, a small towel — it depends of course on who
the respondent is, and what you have asked him to do. But your
research agency will have to calculate an extra ¥300-400 per
respondent into the cost of a regular personal interview survey.

Points to watch

Buying research services in Japan is probably no different from
anywhere else. But the out-of-country buyer needs to watch certain
points that do sometimes get forgotten in Japan.

Find out who will actually do the work, and make sure that they
are up to standard and that their procedures are satisfactory.
Sometimes you will be tempted to buy a well-known international
'name', and forget that they themselves may not have their own
operation in Japan but deal with 'associates' over whose operating
procedures they have little actual control. In such cases, you are
effectively using the local firm, good, bad or indifferent, and its
standards are what you have to worry about. Sometimes even
reputable local firms sub-contract fieldwork to another company
without telling the client. Once again, you end up depending on the
operating procedures of a firm you know nothing about. So insist on
knowing who will be conducting the research and satisfy yourself
about their procedures before you sign the estimate.

Find out what kind of interviewers will actually be involved, in
particular whether or not they are students doing interviews for
pocket-money. In Japan, experience suggests that students are prone
to cheat, and smart enough to conceal it — unless you validate
virtually 100 per cent of the interviews. So unless you are happy with
questionnaires filled in in a coffee shop, make sure the research
agency uses properly trained non-student interviewers.

Ask, too, how the interviewers are being paid. Some firms pay so
much per interview, but this encourages interviewers to skip
difficult questions that need probing, or not to call back as required
before counting someone as impossible to contact. A better way,
used by other firms, is to pay by the hour, thus not penalising an

interviewer financially for continuing to try to contact a respondent, or for taking the time needed to do a good job.

Make sure you understand the sampling procedure the research agency plans to follow. Sometimes the words 'representative sampling' are used more freely than they should be, and can conceal non-representative methods that may not give you the projectable data you need. Through the availability of the government's Local Inhabitants Register as a sampling frame, true representative sampling is perfectly feasible in Japan. It is not always needed, of course, but if that is what you think you are buying, make sure it is what you get.

There are few bargains in Japan. Research costs reflect high salaries and the general cost of living. Payments to interviewers reflect the going rate in competition with other types of work, and increase each year. Expect to pay rather more than in the U.K. perhaps a little less than in France, at least the same as in the U.S. or West Germany. (Of course such generalised comparisons all depend on how costs move in these countries and also on how currency exchange rate relationships change.)

Realistic marketing

Just finding more out about the facts of the market will not solve all your problems. It will not make domestic competition vanish, the distribution system easier to master, or advertising less expensive. But today's marketers are by definition hard-headed people who look for truth and not myth, realistic opportunities and not miracles. And that is where properly conducted research comes in.

It can help you define a gap in the market that powerful domestic competitors may have left unguarded. It can help you find a consumer need that has not yet been satisfied, or develop an emotional or practical appeal that is different — and meaningful. It can help you look ahead, anticipate the trends of the future and help you distinguish between different segments of the market. Homogeneous though Japan certainly is, there are still differences that may offer special opportunities. Some of these differences are demographic — age, region, income, marital status — while others are attitudinal or psychographic. And remember, your product may do well for you even though it only appeals to, say, six per cent of the Japanese population. That is still more people, after all, than the population of Denmark, Ireland, Norway, or Switzerland.

So in planning your efforts in Japan, make full use of your secret weapon — market research. And keep in mind Arthur Hugh Clough's dictum of over 130 years ago — it should be framed on the wall of every marketing man's office — 'Grace is given of God, but knowledge is born in the market.'

10
The Role And Application Of Advertising

DAVID GRIBBIN

Japanese companies making their first advertising forays outside Japan almost invariably begin by running corporate advertisements, or what they fondly believe to be corporate advertisements. Multi-national media with offices in Japan, such as *Fortune* and *Newsweek*, remain a popular choice; Japanese advertisers are placing more and more importance upon local media, hence the lavish promotional exercise staged in Tokyo every year by Gruner & Jahr from West Germany, and the fact that most British publications have their advertising directors drop into Tokyo on the way to Hong Kong. Since corporate advertising, particularly the heavily paternalistic variety favoured by geriatric Japanese company presidents, is disparaged by sophisticated western advertisers and their slick agencies, a western executive may have already formed, or half-formed, the view that Japanese advertising has yet to catch up to the state-of-the-art offerings coming from New York, London or Dusseldorf. When he actually arrives in Japan, and sees domestic advertising, most of which seems to feature an end-of-the-pier duo known as the Pink Lady, he will probably think he's back in the Stone Age, but that's another story.

The enthusiastic development of mass production in the United States gave birth to the advertising industry, which lost no time in telling Henry Ford that 'any colour you like, so long as it's black' was a lousy piece of corporate communication. In the U.S. and Europe, advertising seeks to encourage mass consumers to believe the myth that they are individuals living in an individualistic society far removed from the ultra-conformist societies of the far left and the Far East. This has given rise to an interestingly Orwellian situation: the degree of simulated folksiness and 'character' grafted onto goods and services rises in direct proportion to the way in which those same goods and services become even more interchangeable as the world economy is dominated by fewer and fewer, larger and larger marketers. Casting directors comb the directories, and even the streets, for people with rumpled faces, odd accents and the ability to chomp with rustic abandon on the latest sodium-glutamated facsimile of a Boston baked bean or Italian ice-cream. To be most effective, this process should begin at the earliest possible stage in the evolution of a new product.

This aspect of marketing is well understood in the liquor industry. Glenfiddich malt whisky, for example, owes much of its success to its packaging and the American designer responsible for it recently recounted, with some satisfaction, that a friend of his brought him a bottle of Glenfiddich back from Europe in the hope that he might be able to learn something from this excellent sample of real Scottish packaging. In England, where there's now a fad for old-fashioned red-plush-and-brass pubs, the brewers are paying to have replicas made of the old-fashioned beer pumps they tore out a few years ago when modernisation was the fashion. A few of them are even reviving the rustic names and labels of the small local breweries they bought up to increase their market share and efficiency.

As consumers, Japanese of either sex and all ages judge products in status-role (*bun*) terms. Something which matches his status and enhances his position among his peers in the group will be accorded high perceived value by a Japanese consumer, and he will willingly make considerable sacrifices to attain it.

Although possessions and status-role have always been closely inter-related in Japan, Japanese companies never acknowledge the profit motive as a reason for being in business. Instead, they make public announcements about 'making a better life for all peoples of the world'.

It is no coincidence that the founder of Matsushita Electric, the world's largest manufacturer of electrical appliances and a company famous for its management skills and high profitability, should also be the founder and publisher of PHP (standing for peace, happiness and prosperity), a monthly magazine which is something of a miracle in publishing circles because it manages to combine the high moral tone of the *Reader's Digest* and the circulation of *Penthouse*.

Similarly, it is status, not cash, which makes graduates of Japan's prestige universities compete so fiercely for the privilege of becoming members of the large and powerful Japanese corporations. Once recruited (the ringleaders of the student riots in the late sixties were said to be in great demand because of their proven qualities of leadership) they commit themselves completely to the corporate 'family'. Their goals: to increase their company's market share and their own status as salarymen employed by a company which is a market leader in its field.

In order to build loyalty, most Japanese companies employ elaborate programmes for both leisure and training activities. The former are generously subsidised and the latter include Zen meditation, flower arrangement classes and even special rooms in which employees can vent their frustration by belabouring rubber effigies of the boss of their choice with baseball bats!

In the West, advertising is seen as a most important factor in the constant battle to outsell the competition. But although market share is of crucial importance to a Japanese manager, advertising is far from being the magic ingredient which it so often becomes in the eyes of western marketeers. Since their ability to borrow from the banks depends to a large extent upon their market share, Japanese managers are reluctant to delegate control over such a crucial issue to anything so hit-and-miss as advertising.

Of course, huge sums are invested in advertising, particularly in cosmetics, foods and liquor. Without exception, though, Japan's market leaders in all these fields are companies known for the sophistication and penetration of their distribution systems, as well as their readiness to embrace the latest technology and respond to changes in taste and fashion. The 'real' qualities of the product are felt to be extremely important. Most important of all, Japanese companies compete, at home and abroad, on price.

Since distribution is such a key factor in the Japanese marketing process, it follows that the dealer, rather than the customer, is king; Japanese companies spare no effort to build up dealer loyalty. 'My chairman would never do that in a thousand years,' said a visiting businessman enviously after seeing the chairman of one of the largest liquor companies in Japan roll up his sleeves and bound on to a platform to conduct an impromptu choir of dealers singing the company's latest jingles from its beer commercials. Naturally, dealer loyalty depends, to a large extent, upon the mark-up; Japanese companies make every effort to increase efficiency and enjoy economies of scale because they are caught in a vice. From the consumer side, they are under pressure to reduce the selling price and, from the dealer side, they are under pressure to make the mark-up as high as possible.

The keynote to doing business in Japan is flexibility; success demands not only the willingness to take any one of a number of routes towards the final goal, but also to change the goal itself in the light of changing circumstances. While most producers agree that manufacturing in Japan should be a last resort, the decision as to whether or not to go it alone or with Japanese partners, or to use a trading company or to establish an in-house sales force, must be taken on a case-by-case basis preferably after intensive study of the market.

It is also a good idea to forget what your company does elsewhere. Philips, for example, is known in Europe as one of the giants of the consumer electronics industry. But in Japan, the canny Dutchmen are content to be known, at least for the time being, as makers of electric coffee-makers. They found a hole in the market for this product, aggressively advertised and won over fifty per cent of the market, and are now patiently expanding from this beach-head. They forgot about major appliances such as washing machines, which require an extensive after-sales network, and are concentrating upon electric shavers (scaled-down for the smaller Japanese hand) while preparing to launch hi-fi equipment on the market.

Getting a name known, and building a quality image, calls for mammoth investment in Japan simply because the volume level is so high. Even when a product really does have a unique sales point — as in the case of McVities' biscuits from Britain — it is still necessary to pump huge amounts of time into TV and print advertising. In McVities' case, their Japanese partners are Meiji, a well-established name in the food industry; nevertheless, it has been necessary to invest a great deal of money in prime-time TV to register the product in the minds of Japanese housewives. Other classic success stories, from Coca-Cola to Nescafé, have all been built around massive advertising.

The competition in Japan is too tough for anything but an all-out marketing effort to be successful. Japanese packaging is among the most lavish and creative in the world, with centuries of tradition behind it (see *How to Wrap Five More Eggs,* Hideyuki Oka's classic study published by Weatherhill) and if your product is to sell for more money than the Japanese equivalent it has to look as if it cost more money. Even so, the best packaged and most heavily promoted product will not always be successful. Foreign marketers may take heart — or be reduced to despair — when considering the case of Suntory and the beer market.

Suntory is the nation's largest whisky company and one of the most astute and aggressive marketers in Japan. Ten years or so ago, Suntory, faced with market saturation in its own whisky field, decided to try and get into the beer market, which was dominated by Kirin, which is a part of the vast Mitsubishi group. With their customary boldness and flair, the Suntory people embarked upon a large and ambitious advertising programme on TV, quickly buttressed by new packaging and an up-market foil-wrapped beer clearly inspired by Lowenbrau. Kirin, did not counter-advertise although they did produce their own 'Lowenbrau-type' beer.

After almost ten years, Suntory have captured rather less than ten per cent of the domestic beer market, and although this represents a pretty huge turnover it is, as a return on several billion yen invested in advertising, a dismal picture.

In the fifties, rich American corporations eager to take advantage of a strong dollar, seized the chance to make sizeable direct investments in foreign markets at relatively low cost. To service the resulting business and, long-term, protect the account back in the U.S., the large American agencies followed their clients overseas. Since the wholly-owned foreign advertising agency is only recently possible in Japan, foreigners had no choice but to enter into joint venture agreeements with their Japanese counterparts.

However, unlike the situation in the U.K. where American agencies are predominant, the Japanese advertising industry is very much in the hands of the Japanese. Dentsu, which is the largest advertising agency in the world, is a family business which derives almost all of its income from domestic billings. The second largest agency, Hakuhodo, also derives its income almost exclusively from domestic accounts. Indeed, if foreign and joint venture agencies are judged by market share, it must be said that foreign agencies have yet to make a significant impression in Japan. One reason for this is undoubtedly the facility of the Japanese to adopt the appearance, if not the spirit, of Madison Avenue; another is the incredibly Byzantine, not to say incestuous, relationship existing between Japanese manufacturers and their suppliers.

In Japan there are thousands of small units — production companies, agencies, even advertising laboratories — managing to co-exist with agency in-house creative departments, clients' in-house creative departments and the creative departments maintained by printing companies. The fact that even a highly charged, people-oriented business like advertising can be successfully structured so as to meet the Japanese hunger for harmony, order and group solidarity is, in its way, a kind of minor economic miracle.

With the possible exception of the film business, Japanese advertising is not at all swayed by the expert and his 'baffle-gab'; the Japanese have a healthy distrust of experts of all sorts, possibly because their country abounds in *sensei* (teachers) for everything imaginable. A few years ago, I tried to instil some kind of respect for experts into Japanese clients by delivering a lecture to which I gave the provocative title 'Keeping a dog and barking up the wrong tree'. Having had my copy corrected by people whose English was rudimentary to say the least and my layouts criticised by people whose choice of neckwear proclaimed colour-blindness, I spoke with some feeling. My words were warmly received by the foreigners in the audience and accorded a deafening silence from their Japanese colleagues.

The large Japanese agencies, including Dentsu, started out in life as media brokers, placing ads created by the hot shops and by the clients' in-house creative departments. Between 1955 and 1969, total advertising expenditure in Japan increased more than ten-fold with the result that competition to buy time and space in Japanese media became extremely keen. Even then, media costs were extremely high — between two and four times higher than in the U.K., for instance — and the subsequent appreciation of the yen has made them even higher.

The noise level reaches a crescendo because, particularly in areas like food, pharmaceuticals and cosmetics, Japanese advertisers advertise on a very large scale. In Tokyo there are five commercial TV channels, all colour, all broadcasting from early in the morning (a peak viewing time in Japan) to the small hours of the following morning. Major Japanese advertisers usually make heavy use of all five channels simultaneously, with the result that a would-be foreign competitor has to commit himself to a frighteningly large advertising budget or else retreat up-market where the competition is less fierce, or at least contained within a financial scale less likely to give the home office a heart attack.

Since the heavy cost of promotional activities is yet another reason for trying to find a Japanese distributor and/or licensee manufacturer to bear at least some of the cost, the chances are that the positioning of your product will have been established long before the question of advertising arises. Once the product has been positioned, the advertising approach will be pretty much decided in tandem, though there still remain, of course, several choices to be made. Again, because of the heavy media and distribution costs involved in marketing in Japan, your financial people will certainly be looking for ways to save money without losing impact. Usually, this means trying to adapt an existing TV or print campaign for use in Japan.

When it comes to hidden costs, Japan is the capital of the world. Typesetting, retouching and general mechanical work cost more in Japan than just about anywhere else although photography by comparison is a bargain. This is due to the fact that there are millions of amateur photographers in the country and enough of them decide to turn professional to make competition very severe and prices low. So, in financial terms, creating a new campaign in Japan may be only marginally more expensive than using existing creative work.

Apart from the money aspect, important though it is, there is another reason why the temptation to adapt existing campaigns should be resolutely resisted: everything imported by the Japanese themselves, from religion to Cadillacs, is invariably Japanised to meet local needs and local expectations. (Zen Buddhism is very different from Ch'an Buddhism as originally brought from China, and Cadillacs, though they no longer have the once-obligatory lace curtains at the rear window, are still almost always black and adorned with the gold company badge on the door.)

For centuries, the Japanese people were governed by the authoritarian Tokugawa Shogunate, which followed a policy of almost fanatical isolationism enforced by the most efficient secret police force in history. As a result, foreign products still enjoy in Japan the kind of super-status which can only be conferred by centuries of prohibition. However, the average Japanese consumer has little appreciation of 'real' foreignness and still less regard for the need to develop his sensibilities.

Many years ago, soon after we came to Japan we were entertained to breakfast in the home of an extremely rich and sophisticated Japanese whose taste in matters Japanese is beyond reproach. We were offered eggs for breakfast — a couple of dozen of them, all hard-boiled and stone cold. A well-known food writer recounts ordering 'French toast' in an expensive coffee shop in Toyko — and pouring honey over what turned out to be a kind of Welsh rarebit.

In Tokyo alone, there are an estimated half million restaurants, and they cover the entire spectrum from gourmet French menus prepared by visiting luminaries like Paul Bocuse to tiny 'snacks' serving their own very individual versions of foreign food. It is a mistake to imagine, however, that the Japanese do not know the difference between, say, genuine pizza and 'pizza toast'. They serve the latter because their customers prefer it to the 'real' thing they enjoyed on their Jalpak trip to Rome. As David Ogilvy remarked recently the massive growing traffic between countries makes it unrealistic for a company to think it can advertise in splendid isolation. 'Those days are over,' he said, 'Your corporate image should be consistent everywhere in all countries.'

Ideally, commercials and print campaigns used in other markets should merely be used as a guide for the Japanese creative people. Once they understand the nature of the product/company they should be allowed to develop their own variations on the theme. With the exception of property advertisements, which are rather logical and designed to appeal to a rich élite — doctors and lawyers and politicians — looking for a good investment, almost all Japanese advertisements are aimed at the heart rather than the head, or maybe even a bit lower. As Fumio Yamamoto, manager of the Overseas Advertising Division of Matsushita Electric (Japan's biggest single advertiser) points out: 'The greater part of Japanese advertising is conducted on a lower plane, and aims for a "gut-reaction." So, Japanese advertisements are repetitive, they make frequent use of jingles and . . . often involve testimonials.'

Since nearly half the Japanese population is under 35, and spends its formative years reading extremely rude *manga* comic books, it is not surprising that many commercials should be rather earthy. In Japan, Campari was advertised as offering 'the taste of a virgin' and even in 1979 a commercial for Technics cassette tapes featured an elaborate domino sequence of falling cassettes set in motion by a miniature electrical version of the famous Manneken Piss in Brussels.

Another uniquely Japanese form of advertising to be avoided by the foreign advertiser involves the use of western models who are induced to perform the most ludicrous caricatures of *gaijin* (foreigners) as the Japanese imagine them. This unhappy situation, which dates back to the days when the Dutch traders in Nagasaki were induced to caper before the *shogun*, is compounded by the fact that many of the 'models' are recruited straight off the street. Supermarkets in districts where many foreigners live are happy hunting grounds for the agents, many of whom rely on the language barrier to make off with up to 80 per cent of the model fee!

As has already been mentioned, there is fierce competition for prime media space and time. The foreign advertiser, faced with the need to obtain maximum results for minimum investment due to the dramatic appreciation of the yen, is in competition for space and time with domestic advertisers who may be outspending him by hundreds or even thousands to one.

In my experience, the most effective solution to this problem, and the related one of obtaining the best creative work possible, lies in an à-la-carte approach to the suppliers concerned. This implies that the distributor — who will almost certainly be representing a large number of clients spending in total a great deal of money — should be encouraged to exert his considerable influence with the big agencies and the media, for whom they act as brokers, in order to ensure that good positions and times are obtained. Grouping ad expenditure in key periods usually works better than dribbling it away on small 'invisible' advertising year-round. At the same time, a concerted effort should be made to persuade the big agency to make its services available on a fee basis, leaving you and your distributor free to use, for instance, the agency media and research departments but to go to a separate outside source, with the agency acting as middle man, so as to obtain the best creative work.

By far the most important factor when it comes to evaluating creative work is the supplier's ability to produce ads in Japanese. Whether it is an agency, hot shop, or in-house creative department employed by your distributor you should beware of the so-called *eigo-ya* (English seller) who is a frequent phenomenon in Japanese business. As the name implies, the *eigo-ya* has no expertise other than linguistic ability; it is very easy to find yourself trapped in a situation where the *eigo-ya* interposes himself between you and the management, misrepresenting both parties with gay abandon. Few *eigo-ya* have progressed beyond the classic account exec's 'what time would you like it to be, Sir?' attitude so they would not think of pointing out that if you like the concept or the copy, that is a bonus not an essential requirement. The advertisement is, or should be, designed to sell your product to the Japanese, not to you.

If coordinating a big agency and a hot shop to produce advertisements which you do not understand, but which your distributor tells you are exactly right, all sounds impossibly difficult and complicated, then I'm glad. For marketing in this dynamic market, to an audience of workaholics living in rabbit hutches, as the man from the E.E.C. put it, is horrendously difficult at times;

George Fields, the researcher, and a contributor to this book, who has spent almost three decades in Japan, surely was not exaggerating when he remarked that 'advertising life in Japan is not entirely easy for those schooled in western advertising codes'. But the rewards, in terms of both finance and sense of achievement are, as they say, fully commensurate.

11
The Japanese Housewife: A Marketing Appraisal

GEORGE FIELDS

When talking about the consumer, one is always tempted to do so in terms of the 'average' consumer. Every country has its housewife/ husband/family stereotype, but the usefulness of this 'average' concept depends on the particular characteristics of the society in question. Within a heterogeneous society, such as that of the United States of America, the concept appears almost meaningless; likewise, if there is a great deal of social inequality, and wealth is concentrated amongst a tiny minority, the concept has little relevance in marketing terms.

Japan is neither. It is both a racially and culturally homogeneous society and its wealth is fairly well distributed. Not surprisingly, therefore, the western marketer who goes to Japan is often guilty of sweeping generalisations concerning the Japanese; so it is essential to appreciate from the outset that the image of the so-called 'average' consumer is open to wide interpretation on several levels.

Where the product's future is directly related to socio-cultural factors, statistics on average behaviour patterns can be seriously misleading. For example, the proportion of the weekly budget spent on rice has declined significantly, while consumption of western-type foods has increased noticeably. Since the end of the War the influence of the school meal in which, for institutional reasons, the children were provided with bread rather than rice, has been profound. In one study, we found that, in some cases, the housewife was preparing two types of breakfasts; the traditional Japanese dishes (i.e. rice-centred rather than bread-centred) for her husband and herself and western-type food for the children.

Given this frame of reference, one might assume that the Japanese were ready to take to western breakfast cereals — almost in their stride. Yet for one major manufacturer who invested a considerable amount of money in advertising and promoting a premier breakfast cereal, his product ended up being positioned as a minor snack.

On the other hand, this housewife who went to the trouble of preparing two breakfasts was serving western-style package soups to her children. Hindsight rationalisations are always easy but the soup was more acceptable than cereal simply because it was an extension of the conventional Japanese breakfast which invariably includes salty bean paste soup — and this was in fact being served for the 'other' breakfast to the older members of the family. To many Japanese, the idea of eating cereal with sugar and milk first thing in the morning was equivalent to confronting the westerner with bean paste soup. Clearly, what the children were demanding was a bread-centred breakfast — not necessarily a real western breakfast.

This brings us into the realm of qualitative information input as opposed to quantitative input and in this context we should know that westernisation proceeds along Japanese lines; in other words Japan is seldom able to break away completely from habits and tastes established over centuries. A Japanese housewife, for example, who tells you that her family prefers western meals is, more often than not, referring to a meal that is far removed from the English, American, or French original. Modernisation since the industrial revolution has been synonymous with westernisation; perhaps in Japan we must now recognise that the post-industrial revolution era has started in which modernisation is *not* synonymous with westernisation. In certain areas, modernisation may become synonymous with Japanisation.

An example of the marketing failure of a western food product (developed from a statistical rationalisation) may prove the point even more convincingly. It is a fact that the purchase of shop-baked cakes in Japan accounts for quite a respectable percentage of the food market trade; there was a trend at one stage towards western cakes made from pastry and away from traditional Japanese cakes made from kneaded rice and sweet bean paste. It was also a fact, however, that the Japanese housewife does not bake cakes of any kind in her own home as until recently she lacked a western-style oven. However, the Japanese housewife did make instant puddings from prepared packet mixtures, which was a new culinary trend. Thus, it was argued, if the means were available, surely a substantial market would develop for cake mixes as had been the case in numerous western markets. And suppose, also, one could develop a cake that could be made to an acceptable quality level in the automatic rice cooker — an item that exists in practically every Japanese kitchen? Here, surely, was the basis for creating a new mass market for cake mixes.

Unfortunately, the 'rice cooker cake' was a technical concept, not a marketing concept. Let us look more closely at the reasons why the whole marketing programme failed. First, the rice cooker is in use most of the day — cooking rice; secondly, unlike the oven, the rice cooker was designed solely for the one purpose — given the Japanese penchant of putting things into neat conceptual pockets — i.e. to cook rice; so it may have been too much of a mental switch to consider using the cooker for preparing a completely different type of food. Thirdly, to ask a Japanese housewife to bake a cake in her rice cooker is like asking an English housewife to make coffee in her tea-pot: cooking rice like making tea is a ritual and any 'contamination' of the cooking vessel is instinctively shunned.

But it went further than that: in the traditional Japanese food culture, rice occupies a basic and fundamental position, and young brides used to be judged by their mother-in-law on the way they cooked rice. Although the automatic rice cooker, on the surface, eliminated this problem, the fundamental importance of rice remained unchanged and the rice cooker to this day perhaps cannot be anything but a thing to cook rice.

This case study simply suggests that available 'facts' must, in turn, be interpreted in relation to the social, economic and cultural environment of the consumer.

A few years ago, we surveyed the Japanese housewife's daily chores and examined how she allocated her time. We were fortunate in that we also had similar data for a western culture. Some of the differences were to be expected. For example, she spent less time in cleaning the house, which was simply a function of confined space and lack of furniture; she spent more time washing clothes, which was partly based on the famous Japanese penchant for cleanliness but also again a function of confined space, for it is a lot less tolerable to have soiled clothes lying about in close proximity. Of all the differences, the most important seemed to be the fact that the Japanese housewife spent much more of her time shopping and less time in the kitchen.

In the evenings, a surprising number of Japanese housewives keep a daily book of accounts. The wife allocates her funds skilfully and is in charge of short-term as well as long-term budgeting. In most cases, it is her responsibility to budget for leisure expenses, education, and put aside for the rainy day. She even budgets for her husband's daily pocket money for the average Japanese husband hands over his entire pay-packet to his wife and receives his allocation in return.

The average Japanese family saves in excess of 20 per cent of income, helped by the twice yearly bonus system. This is the highest rate in the world, and is largely through the Japanese housewife's financial expertise rather than her husband's. (The high propensity to save is not a temporary phenomenon and has continued through the worst inflation of the 1970s.)

In the same study we discovered another interesting fact. Many western housewives take a list with them when they go shopping but few Japanese housewives do so. This would seem in contradiction to her budgeting expertise observed which suggested that she is a meticulous planner when it comes to shopping. She is in the sense that she has made up her mind beforehand on approximately how much but not necessarily on what she plans to spend it on at a particular moment.

Her budget planning and lack of a shopping list are not in contradiction because the purchases themselves evolve, *ad hoc*, as she goes on her shopping rounds within the former framework. The emphasis on the Japanese housewife's role *vis-à-vis* the western one is that of 'keeper of the purse,' while perhaps there is more emphasis towards creative functions in the western role.

While some surveys have claimed that the Japanese housewife does not want to spend so much of her time shopping, there is no evidence that this pattern of fresh-foods-first, then other items, is changing. Even with the advent of the supermarket, we have found that the housewife does indeed visit a supermarket but with certain selected product categories in mind and generally also visits other stores on the same day.

The modal family group is four, accounting for over a quarter of all families. The nuclear family — i.e., a couple plus a child or children, and not living with their parents — is increasing steadily and exceeds 40 per cent of the population; three generation families are less than one in five. As total families have increased, the increase has been represented by the growth of nuclear families.

The fragmentation of family units — a modern, post-war trend — has placed enormous pressures on housing. A recent comparison with residential property values — as distinct from commercial property — showed that an equivalent expenditure of about $150,000 in suburban Tokyo and suburban New York would have yielded a house almost four times the size and 20 times the land area for the latter.

In the consumer goods area, women overall are the more dynamic force and influence, with apologies to those who market purely male-oriented products, although even for these, the male's attitude to their product categories must be affected by the revolution that is taking place among the opposite sex.

The pre-war average woman in Japan had five children and spent 19 years of her life in child-rearing, until her youngest entered school. Now that time period has been reduced to less than 10. The average pre-war Japanese woman did not see her youngest child get married. Now she lives a further 24 years. If up to the youngest child's school entry is termed as the 'intensive period of child rearing,' there was less than eight years left after that for the average pre-war woman, but now she has more than 40 years which equals the average life expectancy of the pre-war male.

This trend will continue but at a decelerating rate. A conservative estimate is that about one in four married women is employed at least part-time, excluding those who are in family businesses. About half of all unmarried women are estimated to be employed and many of the remainder are students. Women are thus already activists in society with increasingly important ramifications for the market. They are already demanding a personal say in consumer benefits and their increased confidence is reflected in their demands being personal and not just the demands of spokeswomen for their families.

So far we have only considered the housewife as a consumer in terms of her daily lifestyle; the male consumer, who has relatively little say in the disposal of the daily budget, has also been ignored. On the question of leisure, including the broader spectrum of eating, however, all are participants. Leisure-time spending is now the fastest growing sector in the Japanese market and within it, like any other market in the world, the affluent youth is visible. However, increased leisure time and the frequently asked question of what to do with it may be at the basis of a profound social revolution in Japan. Incredible as it may appear, even at the end of the seventies, the majority of salaried workers do not take up their full entitlement of annual leave. Many still work on Saturdays in Japan, but there has been a concerted effort from some government quarters to make people take more time off because of external pressures from other countries.

Unlike leisure spending on a mass scale, gift-giving has been an established custom in Japan for centuries. Commercially speaking, the twice yearly bonus system (the equivalent of one to six months' pay is given in a lump sum in June and December) has a great impact on the gifts budget. Gifts are given not only to clients, but also to superiors at work or suppliers in order to cement or confirm a relationship. The department stores, which account for 10 per cent of gross retail sales throughout Japan, handle most of the 'gift' sales because of the prestige associated with their name and because the consumer in buying the wrapping projects the image of quality and superiority on to everything he gives — appropriate to the status of the recipient. Daily utility items, for example, are attractively gift-wrapped for the occasion; one of our surveys showed that many homes never buy toilet soap since they always get enough given to them during the two gift seasons to last the whole year.

Even more important for the western marketer, on these two occasions the consumer buys things for himself that he would not normally buy, e.g. Scotch whisky, imported confectionery or prestige brand cosmetics.

★   ★   ★

It is undoubtedly tempting for the visiting marketer to latch on to the obvious similarities he sees in the market place since these give him some sense of security; even so, all too often the marketer's confusion and frustration over what he sees is expressed thus: 'Don't give me that damned rubbish about the Japanese being different. . . they are all people.'

Certainly, the Japanese are the same as anybody else in the sense that they love their children, enjoy their food, and try to provide for comfort in their old age. Stay at any first-class hotel in Tokyo and you will see the same things as you would see in any big city the world over — office buildings, men in business suits, and people drinking coffee and eating western convenience foods. At this level of abstraction — the visible culture — the Japanese are indeed no different from anybody else among advanced industrial societies.

In the final count, however, unless you belong to the lucky few who are gifted with supernatural intuition or luck, then it will certainly pay you to take heed of Japan's invisible culture, which is the world of values, attitudes and motivations. The information input you demand must be qualitative as well as quantitative and, of course, that is what marketing is all about.

12
Marketing Tailpiece: The 'Tranny' And The Fridge

MASAAKI IMAI

In the late 1950s, Sony was having difficulty in marketing its radios in the six prefectures of the Tohoku Region (north-eastern part of the main island of Honshu). Sony radios were sold throughout National Panasonic's retail chains, and it appeared that the sales had reached a plateau and were gradually going down.

Isao Moriguchi, the then product manager of Sony radios, was faced with a need to boost sales in Tohoku. One day, while going over the monthly sales report by a field salesman in Akita Prefecture, he noticed the following footnote: 'I had a visit with a small retailer in Akita and was told that the proprietor had conducted a small experiment on the performance of both National Panasonic radio and Sony radio and found that National's radio performed much better'.

On reviewing this report, Moriguchi called up the salesman and found that the 'experiment' had been conducted by putting the radios in an empty washing machine. He then called up an engineer in the R & D Department and asked whether such an experiment could reveal any difference in the quality and performance of radios. The answer was that it could make some differences depending on how the radio was placed in the washing machine.

The engineer further suggested that, if the experiment had been conducted in a refrigerator, the variables would have been minimal because of the limited space in which to place the radio. He was also advised that such a test would yield no marked difference in performance between radios of different manufacturers but that if the tester was psychologically conditioned to favour a particular model, it was possible that he might feel that such a model was demonstrably superior.

When the engineer was finished, Moriguchi told him to pack a case and go to Haneda Airport, where both of them met and took the next flight to Akita. When they got off the plane at Akita Airport, they were greeted by the field salesman who had prepared the report and together drove to the retailer's shop.

It was well past 7 p.m. when they knocked at the door of the small retailer, and the proprietor, who had no employees, was quite surprised at first, and then felt greatly honoured and privileged to have such visitors come to see him all the way from Tokyo. He apologised for his mischievous experiment which had caused such consternation at Sony head office.

In the meantime, the engineer set out to conduct the test in the washing machine first, and then in the refrigerator, and since the proprietor had been greatly exhilarated by the visit of Sony people, he was, of course, very favourably impressed with the performance of the Sony radio.

After the experiment they retired to the proprietor's living quarters, opened up the *saké* and by the end of the evening they were behaving like bosom friends.

Moriguchi had known beforehand that there would be a convention of all National Panasonic chain-store proprietors in the Tohoku Region within a few days, which was one of the reasons which prompted his visit to this particular shop. Predictably, the visit of the Sony people and the experiment was soon common knowledge among all National chain-store proprietors in the Region.

For a small-time proprietor in Tohoku it was a once-in-a-life-time experience to have someone coming from the Sony head office, an event which he would cherish time and again, and he must have reported the story to everyone he met at the convention. Apparently, the episode brought Sony much closer to many retailers in the area, with the result that sales of Sony radios in six prefectures shot up by about twenty per cent. It is reported that even to this day the proprietor still cherishes the fond memory of Moriguchi's visit and keeps sending in his New Year card. Also, from time to time, he sends in presents, such as the famous Akita *saké* wine.

When Moriguchi, who is now executive director of Brain Inc., a marketing consulting firm, has an occasion to visit Akita, he never fails to call on him and enjoy a drink together.

★ ★ ★

In differentiating Japanese from westerners, it is sometimes pointed out that while the Japanese are emotionally oriented, the westerners are intellectually oriented. Thus, a Japanese manager sometimes finds emotional means much more effective in solving a particular problem. Perhaps, from an 'intellectual' standpoint, the above story may appear to be entirely wild and hard to understand.

For instance, when a subordinate comes and asks for permission to fly to a far-off city together with an engineer to see a small-time retailer who conducted a stupid test in a washing machine and suggests that he is going to conduct the test again, but this time in a refrigerator, an intellectual manager might suggest he puts an ice-pack on his head to cool off.

Whether the above incident made sense or not from the intellectual point of view, the important thing is that it worked. It turned on the people who were involved and turned on the marketing machinery.

In one of Tokyo's suburbs there are two big stores, one is a specialty store selling home appliances, and the other is a general merchandise store. In the spring of 1978, a few days after the former started selling a particular type of electric fan, the latter pushed handbills through the doors of all houses in the vicinity advertising the same electric fan at a price one thousand yen less.

The management of the home appliance store felt greatly obliged to the thirty customers who had bought the fan at his store and with much effort by their salesmen, they located twenty-eight of the original purchasers, apologised for having sold the fan at higher price than their competitor, and returned one thousand yen.

Most customers apparently did not even know that the same fan was sold at a cheaper price but greatly appreciated the 'sincerity' of the store management. It is reported that immediately thereafter these customers made additional purchases at this store averaging ten thousand yen per person. Here again, the Japanese 'emotional' approach did the job.

13
The Japanese And Their Changing Economic Environment*

J.E.W. KIRBY

The 1970s have been a period of great change for Japan. The economy as well as politics and society as a whole are now passing through a transitional period. Two of the major conditioning influences have been Japan's rise to the position of a major world economic power, which has been consolidated during the 1970s, and the so-called oil shock of 1973. Partly as a result of these two developments, the financial system is now undergoing a process of modernisation and liberalisation.

This process has not been embarked on suddenly. It has been under way for some time, but the early 1970s can perhaps be taken as the starting point for a significant change of direction as the Japanese economy grew in strength and began more and more to look outwards. The oil shock and the subsequent slowdown of growth combined with other pressures from outside have added further impetus towards change.

In the 1950s and 60s Japan was first reconstructing and then rapidly developing its industries. Capital was scarce, particularly foreign exchange, and the efforts of the banks, under government guidance, were directed towards supplying industry's needs for working capital and investment funds. The interests of consumers were largely neglected and their everyday needs and even their requirements for housing finance were met more from a high level of personal savings and the convenient system of substantial semi-annual bonuses than from bank or institutional credit. Rapid growth of personal incomes and corporate profits produced buoyant fiscal revenue and, with expenditure for social purposes relatively moderate, financing of the budget presented few problems. On the external side, there was a pronounced cyclical pattern of internal booms leading to over-heating and current account deficits, followed by an adjustment period when, primarily through tight monetary policies, the payments position was restored. There were, at times, significant surpluses but, until the late 1960s, the balance of payments looked vulnerable. Over this period, financial institutions expanded and benefited from the conditions of rapid economic growth if within a closely-regulated framework.

Since the late 1960s the economic environment has changed very significantly. Following the oil shock, Japan's growth aims have been scaled down for the foreseeable future while, in the short term, the world recession, structural weaknesses in certain sectors of the Japanese economy and a general mood of uncertainty felt both by business and consumers have acted as a depressing factor, retarded economic recovery and held back new investment. In the early 1970s, under the Tanaka government, the rate of increase in public expenditure both by central and local governments to meet the increasing demand for improved social welfare services and development of the infrastructure had already begun to outstrip the growth of revenue. Subsequently, the government's efforts to stimulate recovery from the post-oil shock recession through heavy expenditure on public works, combined with the depressing effects on revenue of the slowdown in economic growth and a weakening of corporate profits, accentuated the tendency towards fiscal imbalance and the budget has gone into substantial deficit, financed by large issues of bonds. In the private sector the sharply rising operating costs of the major banks over a number of years, combined more recently with a narrowing of margins due to interest rate movements and the difficulties they have faced in dealing with problem companies have resulted in a decline in their profitability. Externally, the current account of the balance of payments has moved into a situation of large structural surplus.

Trends towards change in the financial system

These developments have led to a number of changes in the way in which the financial system is organised, and to official consideration of the need for further structural reforms. There are four specific trends which might be examined in greater detail: (1) pressures for a freer interest rate system; (2) a shift in the pattern of corporate finance; (3) an increased emphasis in bank and institutional lending on smaller companies and individuals; (4) internationalisation of the yen.

For many years the Japanese financial system has been more

closely regulated than that of probably any other major industrial country. In earlier years, the degree of control exercised was justified mainly by the need to allocate scarce resources (in this case money) in the way best suited to national priorities. Controls of various kinds have also been considered necessary to facilitate the management of the economy generally.

More recently, however, a major motivation for controls and one of the fundamental principles on which the financial system has been organised by the authorities has been that of ensuring the competitive viability of each of the different groups of institutions which make up the system in relation to other groups. Thus, for example, official regulation of the interest rates which may be offered for deposits is designed to protect small banks against large ones and to preserve a place for the former within the system.

Restrictions are placed on the way in which the city banks — the major banks — raise funds, so that they should not compete for the same sources of money as the long-term credit banks and the trust banks. Limitations on the types of securities business which are permitted for banks are partly motivated by considerations of banking prudence, but partly by regard for the relative position of the securities houses. In fact, the division of functions, both at home and overseas, between banks and securities houses has been one of the major internal battlegrounds in recent years.

When change is under consideration, the interests of all parties affected have to be balanced to a much greater extent in Japan than in most other systems. Advance tends to be gradual and based on compromise which, broadly, and over a period of time, seeks to preserve the competitive balance between the different units comprising the system. When one group is enjoying particularly prosperous conditions, it is at its most vulnerable to the granting of concessions to other groups and one can, perhaps, illustrate this by pointing to the permission given to banks in the early 1970s to undertake securities business outside Japan, and the more recent authorisation which they have received for the issue of yen certificates of deposit.

Both of these developments were hotly resisted by the securities houses but, in each case, their voice lost something of its customary power from the fact that conditions in the domestic securities markets were enabling them at the time to earn record profits for themselves.

The greatest threat which is seen to financial stability is excessive competition, and it is paradoxical that in an economy which has become a byword for international competitiveness, totally free competition is only allowed over a fairly limited area in the banking sector.

Liberalisation of interest rates.   Control over the financial
system has revolved round official regulation of almost all interest
rates both for lending (that is for deposits) and for borrowing (on
loans). Over the years, Japanese ministers and officials have spoken
from time to time of the desirability, in principle, of moving
towards a more flexible interest rate structure. Until recently,
despite recommendations from authoritative bodies there has
probably seemed to the Japanese government little compelling need
to make the move in practice.

The pressure of events, however, tends to carry greater weight
than even the most expert advice and developments in financial
markets themselves have confronted the authorities with the need
for a more appropriate interest rate structure.

The Government bond market.   Government bonds in Japan are
allocated in fixed proportions between the members of an
underwriting syndicate which is made up from various categories of
banks (which take up nearly 90 per cent of the total issue), insurance
companies and the securities houses. Terms in the past have been
fixed some time in advance of issue and there has been little
flexibility in pricing. It has been a fundamental aim of the Japanese
Ministry of Finance to borrow cheaply on the government's behalf.
This is an objective which is shared by treasuries throughout the
world. However, whereas in other advanced countries the aim has
generally, if not universally, had to be subordinated to market
conditions and the rates at which investors are prepared to lend to the
governments concerned, in Japan's case, the terms have hitherto
been largely dictated by the government. In these circumstances, it
would probably be true to say that the financial institutions have not
always been enthusiastic takers of government bonds, but considerations of national interest and the advantage to be derived from
coöperation with the authorities have generally ensured that bond
issues have been absorbed relatively smoothly.

Since 1974 there has been a sharp increase in the amount of
government bonds outstanding. The national debt which at the end
of 1972 was ¥11.7 trillion had almost doubled by the end of 1975 to
¥22.8 trillion and by the end of 1978 it had risen to ¥62.3 trillion.
Over the same period holdings of central government bonds by the
banking system rose from 2 per cent of their total assets to 7.8 per
cent. In the summer of 1978, and again in the early months of 1979,
the bond market has shown distinct signs of indigestion. The large
volume of government bonds has depressed prices and raised yields
in the secondary market above those which the government was
offering on new issues. Even taking into account the considerations
of mutual advantage already mentioned, there are limits to the
extent to which financial institutions are prepared to continue to

accept unattractive securities, and they have sought from the government some alleviation of the situation.

The short-term solution has been for the government to raise somewhat the yields on its new issues, to widen the range of maturities available (at present, heavily concentrated on 10-year bonds) and to support market prices to a limited extent by official purchases. The prospects are for a substantial government dependence on bond finance to continue over the next few years, while competing demand from the private sector will grow as investment picks up. Thus, unless a more market-related strategy is adopted, the government is likely to face increasing difficulties in selling its bonds and financing the budget deficit.

The Gensaki market and yen CDs.   Pressures for a freer interest rate system are making themselves felt from other quarters also. One notable development has been the rapid growth of the *Gensaki* market, (literally spot/future) in which bond transactions take place mainly between companies and securities houses on a fixed sale and repurchase basis, for maturities of up to one year. Unofficial estimates are that the size of the market has increased from the equivalent of $5 bn. in 1976 to some $20 bn. at present and, until the Bank of Japan began to take measures in 1978 to free interest rates in the call money market and the discount market, it was generally believed that this was the only financial market in Tokyo where rates were free from official influence. This has been its particular attraction for investors. The market has flourished in a period of uncertainty about the future outlook for business, when companies have held back plans for new equipment investment, looked for financial opportunities for the employment of their surplus funds and seen no alternative short-term instrument offering comparable yields to those available in the *Gensaki* market.

Such funds would previously have tended to remain on time deposit with a city bank and this loss of business combined with their inability to compete on interest rates and flexibility of terms with the *Gensaki* market was a major factor behind the banks' call to the authorities to be allowed to issue yen Certificates of Deposit. At this point, considerations of the competitive balance mentioned earlier began to arise. The securities houses, a powerful lobby, saw a potential loss of business for themselves if the city banks were allowed to offer CDs. The smaller commercial banks, and the long-term credit and trust banks saw the possibility of unwelcome competition for their own sources of funds. All made their views heard and the Ministry of Finance, the final arbiter, has recently declared the outcome. Banks will be allowed to issue Certificates of Deposit but to soften the impact on other institutions restrictions have been placed on their volume, their maturity (limited to six months' maximum) and their negotiability. Interest rates are, however, free.

The significance of the trend for freer interest rates is that decontrol in any individual sector of the market is likely to create pressures for decontrol in other sectors, if distortions are not to arise. Thus liberalisation in short-term markets is likely to have effects, ultimately, in the market for government securities themselves.

Shifting patterns in corporate finance

The prevailing pattern of corporate finance in the post-war period has been a very high dependence on bank borrowing, with a correspondingly low rate of direct financing from internal funds or the capital market. There are signs of a change in this pattern, with, for instance, a sharp rise over the past three years in domestic and overseas security issues by Japanese companies, while their borrowing from banks has stood still or even fallen. There is doubt, however, as to whether the shift is a cyclical one which will reverse itself in due course, or a structural change. Figures probably overstate the fundamental move away from the banks but a trend seems to be under way towards a diversification of sources of fund-raising and sounder balance sheets by western standards, which will become even more prominent in the future.

The single most important influence on corporate attitudes has been the extent and the duration of the recent recession. While previous downturns have placed strains on corporate finances over a relatively short period, the pressure caused by the recession which followed the oil shock have been very much more severe and have demonstrated vividly — through the monthly bankruptcy figures and the number of enterprises requiring financial relief of one kind or another — the dangers of a heavy burden of interest payments at a time when profits are no longer growing.

While the lessons derived from this experience relate to a company's ability to survive under difficult conditions, other motives have contributed to a change in attitudes, perhaps having its origin well before the crisis period of the oil shock. As major Japanese companies, particularly those in export sectors, grew under the boom conditions for export markets of the late 1960s and early 1970s and established international reputations, new financing opportunities opened up for them, as well as possibilities not only of gaining greater independence and flexibility by diversifying their financial structure, but also of improving profitability by covering foreign exchange risks and taking advantage of interest rate differentials.

Fund management by international companies worldwide has

become increasingly sophisticated in recent years, and there is no evidence that Japanese corporate treasurers have been slower to learn than their foreign competitors. The establishment of overseas subsidiaries by many Japanese companies has also encouraged increasing recourse to international capital markets to finance their activities.

If the analysis is correct that a shift away from the banks has begun and is likely to gain strength, the implications for the Japanese economy are very significant. Although none of these should be exaggerated, they would include a weakening of the enormously influential position which the banks have occupied in the Japanese system in the post-war period, the further development of the domestic capital market, and greater difficulties for the authorities in their domestic monetary management. This has derived much of its effectiveness in the past from the extent of the dependence of the Japanese system on a pyramid of credit, which in turn has given the authorities considerable and rapidly-acting control over the level of economic activity, by their power to vary the supply of credit through the window guidance system.

Consumer credit

There has been mention of the neglect in earlier years of consumer credit which, in Japanese terms, includes also mortgage finance. The prospect that, in the period of reduced growth which seems to be ahead, the financing requirements of their major customers are also likely to be reduced, has caused the city banks to rethink their lending strategies. Their own declining profitability has been another consideration as has the move towards other sources of finance than bank borrowing. For all these reasons, the city banks are vigorously seeking new business from small and medium companies, with potential for rapid growth, which have previously been the clients mainly of regional banks. They are also placing increased emphasis on consumer finance. This is seen clearly in the figures of bank lending for housing which increased between 1970 and 1978 from 1.7 per cent of total bank credit to 9.3 per cent.

It is surprising that consumer finance for other purposes has not increased proportionately at all over the same period, accounting for only some 0.6 per cent of total bank credit. Possibly figures for credit card transactions and loans to consumers by finance companies, in both of which banks have a sizeable stake, would show a different picture. Although it has been said that the Japanese prefer to pay cash and are not attracted by consumer credit of the type which is so common in the West, the rapid growth of a department store like Marui which trades primarily on credit would seem to contradict this theory. It certainly appears to be the case that consumer finance, even if for the moment it still plays a very minor role, is a growth area for the future.

Internationalisation of the yen

Internationalisation of the yen is rather an imprecise term, but it is a convenient one to describe the expansion of the international activities of the Japanese banks, the increasing use of the yen as a reserve currency and the development of Tokyo as an international financial centre.

Internationalisation of the yen springs first from the strength of the currency itself, demonstrated particularly by the recent performance of the Japanese economy; secondly it derives from a change of attitudes on the part of the Japanese authorities themselves. This has been gradual rather than sudden, but perhaps the statement made by Vice-Minister Yoshida in January 1976 that the Japanese government was no longer going to resist the holding of the yen as a reserve asset by foreign governments represented a major turning point. This was a long way from saying that Japanese authorities welcomed the role of a reserve currency for the yen, but since that time their basic attitude seems to have been one of acceptance of the inevitability that the yen is going to play an increasingly important part in the international monetary system.

The process of internationalisation has been hastened by the recent emergence of huge current account surpluses, the consequent strengthening of confidence in the yen, reflected in the rise in its external value, and the encouragement which the Japanese government has given to the export of capital to offset the current account imbalance.

The opening up of the Tokyo domestic market to foreign official borrowers has been one of the most striking aspects of this development. In 1978 yen-denominated bond issues by foreign official borrowers amounted to the equivalent of $3.6 bn., putting Tokyo into third position as a world capital market behind New York and Zurich. 1979 has seen a further significant development, with the first issue by a foreign corporation, Sears Roebuck.

Japanese banks too, which first began to show their muscle in international lending in 1973, emerged in 1978 as a major force, with syndicated yen credits to overseas borrowers amounting to some $3 bn. and dollar credits of $9.2 bn.

Loans denominated in yen, increased Japanese aid flows and pressure from her trading partners to invoice their exports in yen will all tend to increase non-resident holdings of yen and to promote the growth of Tokyo as a financial centre. Until recently, official yen holdings have been fairly small. Up to 1976 they did not much exceed $500 m. in aggregate and in 1975/76 represented only 0.6 per cent of total world reserves. Since late 1977 however, there has been a rapid build-up of non-resident free yen balances, which rose from around $2 bn. in October 1977 to $ 4.3 bn. by March 1978. Unofficial estimates are that the 1979 total is now some $5.8 bn. and all the indications are that it is not only private balances which have been increasing. These figures are quite separate from non-resident holdings of Japanese bonds and equities.

Apart from the implications for the domestic banking system, the issue of yen CDs represents another important addition to the facilities which Tokyo offers as a financial centre, and the prospects over time are for the yen to assume an increasingly significant role both as a reserve and a trading currency and for Tokyo to make further advances as a major international financial centre.

★   ★   ★

The monetary developments described above have taken place against the background of an evolving fiscal situation which is worth a brief examination in conclusion.

In the 1970s, there has been a major change in the orientation of fiscal policy. In the first place, there has been a switch in emphasis from the primary objective of the 1950s and 1960s (to stimulate rapid economic growth by promoting the expansion of manufacturing industry) to an aim of improving social welfare, the infrastructure and the general living environment. This change in aims has partly come about as a response to new demands from the people themselves, but partly also because of the emergence of various constraints from both the internal and external environment on a continuation of the previous pattern of economic development.

The most significant economic consequences of these new social imperatives have been the need to increase public expenditure and to raise resources to finance this increase. The recession, following the 1973 oil shock, while temporarily inserting different priorities — that of stimulating general economic recovery, at the expense of a temporary scaling down of social welfare and other expenditure — has brought the problem of public finances even more clearly into the open. The government and people have had to become accustomed to levels of deficit financing which have caused unease both at the practical and psychological level. Over the next few years, taxation will have to be increased to restore fiscal balance, and a major element in this will be the planned introduction of a general consumption tax . Even so, there will be an interim period, at least, where a heavy dependence on bond financing will persist. It has been argued above that the imperative of selling government debt in large quantities is one of the factors which is likely to act as a catalyst over the next few years in producing a freer interest rate structure throughout the financial system.

In some ways, this will make monetary management less complicated, but it will also create the new challenge of regulating a freer economy which is likely to be less responsive to official signals than in the past. Whereas, looking at the broad spectrum, over most of the post-war period, a major concern of monetary management has been to strike a balance between the needs of manufacturing industry and the private sector, with the government's demands fairly modest, in the future the balance will have to be struck in rather different proportions with the government playing a much greater part.

One could say that Japan's industry has developed and modernised at a more rapid rate than the financial system which has fed it. The financial system which in the early post-war years was designed to foster and protect the growth of a still immature and vulnerable economy is now, itself, developing into a form which is more suited to the characteristics of today's fully mature Japanese economy and the major role it has to play in the international system.

* This chapter is based on a paper given to the British Association of Japanese Studies in April 1979.

14
Foreign Banks In Japan

JOHN W. ROBINSON

The business of the foreign bank in Japan is wholesale-oriented and is controlled by the following regulatory constraints.

    a.   Foreign banks are not licensed to operate as savings banks or trust banks and cannot therefore undertake such business.

    b.   With respect to foreign currency loans, (Impact Loans), facilities for less than one year are not permitted.

    c.   With respect to conversion of foreign currencies into yen and accepting convertible free-yen deposits from their branches, foreign banks' operations are restricted within certain limits by means of a Swap Quota which is allocated to each foreign bank by the Japanese monetary authorities.

The quotas differ from bank to bank and whilst the *individual* allocations are not published, factors which presumably influence the size of the individual bank's quota include (i) length of operation (i.e. how long the particular foreign bank branch has been established in Japan), (ii) significance in the market (i.e. whether or not the bank fully utilises the limit imposed upon it — when an increase is being considered by the authorities), (iii) degree of observance of the written and unwritten regulations.

In the light of these restrictions foreign banks' operations may be classified under three major and two minor headings.

Major categories

Impact loans.   Medium-term foreign currency (mainly U.S. dollars) lendings to major Japanese corporations in Japan with interest being quoted and charged as a margin over the London Inter Bank Offer Rate, (LIBOR).

Short-term yen loans.   Working funds extended to Japan-based corporations (both Japanese and foreign). Such loans are usually for three months with the interest rate being quoted as a margin over the Bill Discount Rate. These loans tend to be rolled over quarterly with the interest rate being adjusted up or down at rollover in line with movements in the Bill Discount Rate.

Foreign exchange.   The purchase and sale of foreign currencies in connection with their commercial business and branch funding under their Swap Quota. The latter necessitates borrowing U.S. dollars for the period chosen by the bank, selling them spot in exchange for yen and entering into a forward contract to buy back the dollars with yen on a given date in the future, this date to coincide with the maturity of the U.S. dollar borrowing, when running a fully matched book.

As it is a requirement of the Japanese monetary authorities that all dollar or other foreign currency borrowings which are swapped into yen must be covered by forward purchase contracts except to the extent of the Bank of Japan open position limit imposed on each bank, there is a constant need by all foreign banks to undertake such foreign exchange business to fund their operations.

Minor categories

Discounting of commercial bills.   Normally under letters of credit advised or confirmed by them. More often than not the discounting is done by the beneficiary's main Japanese bank and the bills and

supporting documents are passed to the foreign bank for reimbursement only. In this way the Japanese bank provides the finance and
therefore receives the major portion of the earnings with the foreign
bank receiving a nominal handling charge for processing the
reimbursement.

Establishing/advising/confirming letters of credit.   In view of
Japan's concentration on exports foreign banks in Japan are much
more involved in advising and confirming letters of credit opened
through them by other branches of their bank, group members or
foreign correspondent banks than they are in establishing letters of
credit in favour of foreign beneficiaries.

Funding

In aggregate, excluding contingent liabilities, 53 per cent of foreign
banks' funding is accounted for by borrowings from the relative
Head Offices and overseas branches (i.e. outside Japan) with 26 per
cent being provided by deposits and 1 per cent being raised through
the Tokyo Bill Discount Market. In the latter category foreign banks
discount their own paper through the brokers for periods up to three
months. The market beyond three months is extremely thin.

Growth of foreign bank lendings

Turning now to the major application of the funds raised, of the 55
foreign banks operating branches in Japan (June 1977) there are two
quite distinct groups.

The first group comprises the three largest American banks —
Bank of America, Citibank and Chase, the smallest of which is, in
asset terms, more than three times the size of its nearest competitor.
These three banks as stated above account for 36 per cent of total
foreign bank loans with Citibank alone taking 14 per cent.

The second group comprises 52 banks which range in size of loan
portfolio from a low of ¥5.3 billion to a high of ¥176.2 billion.

However, the halcyon days of dramatic growth are over and with
low domestic demand and downward pressure on margins it will be
extremely difficult to sustain the recent level of lending, let alone
expand it.

Foreign exchange

With Japan's projected slower growth, foreign banks must look to
alternative ways of servicing Japanese corporations. As a consequ-

ence of the internationalisation of Japan, there will be a tendency for corporations to move, albeit slowly, away from their historical ties to the most competitive sources of funds or services and foreign banks must be prepared to meet these needs. Off-balance sheet services such as foreign exchange is one area in which truly international dealing banks have a decided advantage. However, development of this area depends on the Japanese authorities since the Tokyo market is still very much local in nature.

Foreign banks which have set up fully-fledged dealing operations as opposed to foreign exchange covering/funding departments are becoming a significant force in the market but the Bank of Tokyo, as the only authorised specialised foreign exchange bank in Japan, still maintains an almost monopolistic hold on the Japanese foreign exchange market. Nevertheless, with the internationalisation of the yen and its possible use for third country trade and provided some of the deficiencies mentioned above are corrected in part at least, then there is scope for growth.

One of the major obstacles for foreign banks in setting up is the lack of available trained international dealing staff. In Japan's very traditional labour market it is not the custom to move from one employer to another except at or around the retirement age of 55. All foreign banks benefit from this system but since foreign exchange dealing is relatively speaking a young person's game, it is necessary at least initially to import trained expatriate dealers and back-up staff to set up such an operation and to train locally recruited staff straight out of university. This is expensive and not at all easy since the principle of making instantaneous decisions without prior group discussion is alien to the Japanese character. However, this is not insurmountable and those banks which have started dealing operations seem to be overcoming quite successfully what at first sight seemed an almost impossible task.

Those foreign banks most actively engaged in dealing in Tokyo may be split into two groups. Those long-established and having a large commercial base from which to operate include (in alphabetical order) Algemene Bank, Chase Manhattan and Morgan Guaranty. The second group, which with the exception of one name is relatively new but is fast becoming significant in the market includes (in alphabetical order) the following banks: American Express International Banking Corp., Barclays Bank International Ltd., Union Bank of Switzerland and Westdeutsche Landesbank.

In view of the worldwide activities of Japanese corporations, the liaison role which takes up much management time in Japan is of most importance in this the land of personal relationships, and whilst the prospects for new branches may not be particularly bright, at least along traditional lines, international banks cannot afford to be without a representative in Japan.

15
Services Of A Japanese Bank

KUNIHIKO KOBAYASHI & TAKASHI SUGIYAMA

This is an introductory guide to the various facilities any major Japanese bank can offer foreigners who wish to start or enlarge business in Japan.

As most leading international banks now have offices in Japan, you will most likely find your 'home bank' operating there and can use its facilities in your Japanese business. You will eventually discover, however, that Japanese banks can offer a wider range of services. In simple terms, Japanese banks are more conversant with the local market and have all the services you need at their fingertips.

Mitsubishi Bank, for example, one of Japan's largest banks, maintains 202 branches at major industrial and commercial centres throughout Japan and transacts business with about 60,000 enterprises covering every sector of the Japanese economy. Backed by authoritative information gathered through these transactions, the bank's response to the most complex customer requirements can be swift and effective.

Typical services offered include the following:

For businessmen who wish to export new merchandise to Japan:

- To supply detailed information on the Japanese market regarding their merchandise.
- To introduce Japanese wholesalers and retailers.
- To arrange such operations.

For businessmen who plan to set up a joint venture with a Japanese firm or enter the Japanese market by taking over a Japanese firm:

- To provide a list of Japanese firms wishing to do business with a foreign firm.
- To conduct credit and field research on these firms.
- To introduce or arrange prospective deals.

For businessmen who are interested in Japanese technology or merchandise development:

● To supply up-to-the-minute information on unique technology or new products of Japanese firms.

For businessmen who wish to invite Japanese firms to their country to establish a joint venture with a Japanese firm there, or to tie up with a Japanese firm by concluding a technological assistance or agency contract:

● To supply a list of Japanese firms who plan to enter their market or wish to purchase foreign technology.

● To conduct credit and field research on these firms.

● To introduce or arrange promising deals.

Most major banks have an information and business development office which can be depended on as a reliable source of information on Japanese enterprises, industries, markets and all the necessary legal procedures.

Japanese banks also offer certain advantages in looking after the financial affairs of a foreign businessman or company. They are able to provide more effective assistance with yen requirements either by extending loans or by accepting deposits. Private customers enjoy cash dispenser services or standing order services for monthly utility bills, details of which are given later. Other services include credit cards, leasing, factoring and computer data processing provided by affiliate companies. Experienced English speaking bank staff, of course, are located in all main branches and at the Overseas Business Division of the Head Office, with specialists on hand to promote business in Japan with foreign enterprises.

Considered in more detail, the main services offered are as follows.

Deposits

As a foreign resident in Japan you will doubtless be aware of certain differences between banking in Japan and in your own country. Not only can the language barrier complicate the comparatively simple procedures for opening an account, but also the reluctance to accept payment by cheque can be frustrating when you are shopping. Communications problems also prevent many foreign residents from making the best use of their accounts, so that they tend to use the local branch of a bank from their home country, not realising that the services such a bank can offer are sometimes very limited.

Opening an ordinary deposit account at any branch of one of the

leading Japanese banks is a very simple procedure. There are staff
who can speak English, and you will not normally be required to
give full references. You are free to base all transactions upon your
personal signature instead of the normal Japanese method of using a
hanko (seal) if this is what you wish.

While an ordinary deposit account is not a cheque account, it does
pay interest on the daily minimum balance, and higher rates of
interest are available for time deposits, which can be entered in the
same passbook. This kind of account can be used for bank-to-bank
payments, and for standing orders to pay your utility and other
regular bills.

This kind of account also eliminates much of the inconvenience
felt by those who cannot use cheques in Japan as widely as they
would at home. By requesting a cash card, you will be able to
withdraw, say, ¥100,000 at a time from not only your bank's cash
dispenser machines, but also the inter-bank cash dispensers often
located in department stores, busy shopping areas and subway
stations.

Below are details of the various kinds of accounts that can be
opened in your name. The interest rates on ordinary yen deposits are
as follows, as of 1 June 1979.

	Term	Interest rates %p.a.*	Note
Time Deposit	2 years	5.50	
"	1 year	5.25	
"	6 months	4.50	
"	3 months	3.25	
Deposit at notice	over 1 week	1.75	
Savings Deposit Account	Withdrawable at anytime	1.50	A withdrawal slip is used for withdrawals
Current Account	"	—	Cheque books in English are available

*In certain cases, 15% withholding tax is charged on the interest.

Non-resident depositors, including those with diplomatic status,
may open free yen deposit and current accounts. There are also U.S.
Dollar and German Mark deposit accounts. Rates, of course,
fluctuate in accordance with the international money market and
interest rates on deposits of over $100,000 are subject to negotiation
and are usually higher.

Loans

Japanese banks usually extend credit facilities to customers either in the form of discounts of commercial bills, loans on bills, or loans on deeds. Overdraft facilities are also available but these are not so prevalent in Japan as elsewhere.

Banks discount for vendors of merchandise the commercial bills they receive from the purchasers as sales proceed. As this is done on a 'with recourse' basis, banks can usually rely on both the credit of the vendor — the bank's client, and the purchaser — the issuer of the commercial bills. Banks may sell and liquidate such commercial bills in the inter-bank market whenever necessary. For the above reasons, banks usually prefer this form of credit facility to the under-mentioned.

Loans are granted to borrowers on bills issued to banks or on deeds submitted to banks. This covers short term or middle and long term financing. As banks regard this as less creditable than commercial bills discounts, the borrowers' credit standing is more severely scrutinised.

Overdrafts, as stated earlier, are not strongly favoured by Japanese banks. The main reason is that once an overdraft credit line is established, banks cannot thereafter exert effective quantity and quality control of the actual loan balance. Quantitative control of the actual loan balance sometimes becomes necessary under the so-called 'window guidance' of the Bank of Japan, especially during a tight money period. 'Window guidance' is the means by which the Bank of Japan controls credit on individual commercial banks, under which a quarterly ceiling is set for each bank for the amount of credit increase during the quarter.

Overseas Business

Assistance with a foreign firm's import and export business, including the transfer of funds, is also available.

The Mitsubishi Bank's current international facilities, for example, include seven branches, 10 representative offices, four subsidiaries and more than 20 affiliates. Through this network there are correspondent arrangements with approximately 3,000 offices of 1,000 banks in 133 countries.

Other services

Local money transfer service. In Japan, bank transfers are commonly used as a means of settlement between business enterprises.

Cash dispenser service.   Upon request a cash dispenser card can be issued to savings account (resident yen only) depositors. By using their personal 4 figure code, card-holders may withdraw funds from their savings accounts. Dispenser machines operate from 8.45 a.m. to 6.00 p.m. on weekdays and from 9.00 a.m. to 2.00 p.m. on Saturdays. Bank windows close at 3.00 p.m. on weekdays and after 12.00 noon on Saturdays. Card-holders can also withdraw cash from their savings account at any of their bank's branch offices in Japan.

Standing order service.   A standing order service exists at all banks for monthly payments of gas, water, electricity and telephone bills.

Safety deposit service.   Certain banks offer a safety bag service which is like a mini safety deposit box but is far more economical.

Household matters.   Finally, major banks welcome enquiries on such household matters as finding a residence as well as other matters relating to your personal life in Japan.

16
Stocks, Securities And The Brokerage Market

SIMON GROVE

Japan is still definitely a capitalist and free enterprise society and any reference to the national consensus indicates that it is likely to remain so. But there are subtle differences, as is true of so much else, between the foreign and the native versions of what are broadly identical economic systems. We can analyse this by reference to the history of economic thought — Japan has produced seminal writers on the subject since at least the late eighteenth century — or by examining the bureaucracy, the banking system, the taxation structure, or in many other ways, but perhaps the most direct, interesting, and even profitable way is to study the Japanese stock market.

There are several stock markets in Japan, still nationally independent of each other, but the one that matters is the first section of the Tokyo Stock Exchange. In fact, there is considerable overlap in that many if not most shares listed on the TSE first section are also listed on such exchanges as Osaka, Nagoya, Hiroshima, Fukuoka, Kyoto, Sapporo, or Niigata. The converse is not true; many companies are only listed on their regional exchange. There are second sections in Tokyo, Osaka and Nagoya where the securities of smaller capital companies or less actively marketable stocks are traded; listing requirements are also less exacting. There is also an 'over the counter' market where it is possible to trade in a variety of unlisted securities.

The Tokyo Stock Exchange is located at Kabutocho, near Nihonbashi. Here, in 1977, some 83 member firms dealing in 1451 equities transacted a daily average of $377 million (in New York the figures were $596 million and 2177 equities; in London $78 million and 4,475 equities). Tokyo, therefore, is one of the three great stock market centres of the world. What is more, the Tokyo markets are great not merely on the grounds of size but because they are easy to trade in — what you can buy you can expect to sell at a market price which might be more or less than you paid, but at least the market is active enough to give you a good chance of both getting in and getting out.

In Japan there are no shares so priced that to buy just one would cost more than an average working man earns in ten years, as is the case in some continental European stocks. There is diversity — you can if you wish invest in a wide range of industries or sub-industries — more, in fact, than in the U.K. because takeovers are more or less forbidden — and even have a choice of several companies. It is easy to get information, easy to place orders, and easy to get your shares when you have paid for them or your money when you have sold them. It is easy to make contact with stockbrokers — in Japan it is often only too hard to avoid meeting them as door-to-door salesmen are employed and the securities houses have street-corner branches everywhere.

In each of the world's three dominant markets most of these facts are broadly true; in Japan, the combination of such factors with a well-educated, well-informed, wealth-motivated, speculatively in-clined population that participates in a high savings rate and enjoys a tax structure designed to motivate such people has produced and is producing a 'peoples' capitalism'. Admittedly, shortly after the post-war reorganisation of the securities' market, private individuals owned over 60 per cent, whereas the figure is now just over 30 per cent, but during the great economic boom of the last 20 years when the number of shares has risen ninefold, the total held by individuals rose sixfold, and this during the period when institutions such as banks and insurance companies have been expanding and simultaneously playing a greater role in the market. In 1950 there were only 4.5 million Japanese individually owning shares; today there are nearly 20 million.

As noted above, the Japanese have a high savings rate — over 20 per cent of their salaries — partly due to the fact that until recently social welfare has been fragmentary (though this is now growing as vigorously as Japanese economic organisms always seem to do), but also due to an instinctive belief in the merits of saving. The most recently published figures — end of 1977 — show that the average household in Japan had an average income of 3.8 million yen and an average of 4.2 million yen in savings. If we eliminate the 786,000 yen shown as the average saved in life insurance — presumably at current cash-in value — we find average savings come down to 2,983,000 yen of which 765,000 yen — over 25 per cent — is in securities. In practice, two-thirds of this is in bonds — a popular form of saving in Japan — and the bulk of the remainder of savings is split between time deposits with banks and ordinary savings accounts. So there is still plenty of private cash which could go into the stock market.

The securities houses, or stockbrokers as we would call them, have branches in most major shopping areas and anyone can go in and gaze at the 'big board' with price changes showing up minute by minute. Unlike a British brokerage house, the sales personnel (many of them girls) in Japan sit at a counter and wait for the customer to come to them.

There are 256 such brokerage companies in Japan, employing a total of nearly 90,000 people. Some of them are very big indeed, and the larger ones are quoted companies themselves. The majors, such as Nikko, Yamaichi and Daiwa dominate the industry and as names are just as familiar to the Japanese as Toyota, Sony or Yashica. As they spread their activities overseas they will become as well known to people in Great Britain and the U.S.A.

Securities are bought and sold through brokers, but there is no jobbing system as in London. Instead, offers to buy and sell any particular share at a stated price are written down on sheets of paper kept at trading posts on the floor by clerks called *saitori*.

Whatever your views are on stock markets, it is hard to ignore an industry with an annual turnover of well over 100 billion dollars. This is not only a greater sum than the total British Budget, but is also comfortably more than the Gross National Products of a great many national economies. Admittedly, such comparisons are unreal as stock exchange turnover generally, and more in Tokyo than anywhere else, represents the same money moving to and fro to the wonder and entertainment of all involved, some profiting and some losing thereby. Nonetheless, it is an awesome sum, with implicit horror ahead if everything went wrong at any time. What purpose does it all serve?

The customary justification for stock exchanges is as a cheap source of fresh capital to help companies to expand to the benefit of owners and workers alike. Although this may be the rationale of the Japanese securities market, it is hardly the main reason. An examination of the situation leads one to an understanding of the nature of Japanese post-war capitalism which has worked so well for the Japanese — bringing them from the ruins of defeat to their current economic position in a generation.

In the West, equity or ownership is central to all businesses including joint-stock companies. Ownership confers rights as well as bringing risks, and normally investors expect a good return on their 'risk capital', which they gear by as much as 50 per cent borrowings, creating an average debt-equity ratio of 30-70. The role of management, and the aim of business, is to maximise efficiency and profits and thereby secure over a period of time an optimum return to shareholders. If business is poor, or the outlook bad, the directors will reduce their overheads, including the number of people employed, as a matter of course, or face removal for mismanagement of the assets with which they have been entrusted.

If a company is making idle use of assets, returns will be poor, people will not regard the stock as a good investment, and the price will fall to substantially below net asset value. In such cases, entrepreneurs emerge who 'take over' the asset-rich company, often in return for newly-printed shares which effectively means buying out the owners with their own money. By rationalisation, including closing underemployed plant, and sale of assets, they can realise the undervalued assets in the form of profits, and this has been considered desirable as a spur to management or a stimulus to the economy.

In the case of every statement made above, and many more that could be made, the exact reverse is true in Japan. Without spelling out every point, the following are particularly important.

Takeovers and mergers, except in the case of bankruptcy, or near-insolvency, or industrial rationalisation in recession, are all but unheard of in Japan as far as quoted companies are concerned. This is chiefly due to the group ethos which, applied to companies, means that there are corporate relationships with banks and probably with other members of one of the major loose industrial groupings which do not permit the outside predator to carry out his raids.

An examination of stockholder lists shows an average of 30 per cent of shares of listed companies in 'safe hands'. At least one bank, and probably more, will feature as a major holder of up to 10 per cent of a company's stock (the maximum figure for any one bank and any one company). The banks will grant long-term loans and shorter debt on a loan basis against a compensating balance of up to 50 per cent.

Another aspect of the group mentality is that if mergers occur — as they do between unquoted companies on their way to growth and listing — the consequent factionalism in the new company between members of the constituent parts makes management even harder than is normally the case through factionalism between followers of various leadership lines in totally unmerged companies. In the case of a takeover brought about by bankruptcy, those of the 'defeated' party will never secure equal treatment or esteem with the 'victors', which again leads to disharmony.

The objective of a Japanese *kabushiki kaisha* (k.k.) is more than merely to limit liability to raise capital and to secure ownership in shareholding form. It is also very much removed from the 'maximise profit and dividend' theory cited above. Immediate yield becomes meaningless in a growth-oriented society.

The implicit objectives of a k.k. are to form a group to cooperate in an enterprise, aiming to increase output, profit and market share and thereby to raise the standard of living of the inner group (the workers) the outer groups (shareholders, the neighbourhood, etc.) and ultimately of the supreme group, the people of Japan. It is this motivation which has made Japan successful, and it is only a modification of traditional Japanese hopes. This explains why young leftist activists at university find nothing incongruous in subsequently joining large companies. Companies, like people, in Japan are all greater or smaller parts of groups which operate on the basis of long-established relationships. This factor is central to the whole topic of this book — doing business in Japan. Paradoxically, the one area of Japanese life where the constant switching of loyalties is not only admired but necessary is the stock exchange. You can't work for Hitachi one week, Toshiba the next, but you can buy and sell their stocks with impunity and profit.

The shareholder, like everyone else, has his role. It is not to suck a large proportion of the profits out in the form of dividends. Less than half of profits earned are paid out in dividends, and these profits themselves only represent one-twelfth of capital initially invested. The dividends paid total less than one twenty-fifth of the total market value ('share price') of all shares. Put most simply (and roughly) the average Japanese share has a par value of ¥50, a price of ¥250 and earnings of ¥10 on which it pays out ¥5 (these figures are, in fact, very close to the broad majority of shares).

But the shareholder does not complain. To begin with, his objective is unlikely to be income, which is taxed at source, but capital gain on which, unless he is a very big operator indeed, he is excused or can evade tax. In any case, reflecting the philosophy outlined above, he can hardly claim to be the owner of 'his' company. His share in the company amounts to only a fraction of the debts, including loan finance, of the company. In other words, the money the company uses is not so much what it has obtained from him as from other sources. In international terms, the comparisons are even more striking.

Total assets

	Shareholders equity, %	Liabilities, %
Japan (1976-7)	13.7	86.3
U.S.A. (1976-7)	53.8	42.6
U.K. (1974-5)	40.0	60.0
West Germany (1975-6)	28.2	71.8

The Japanese companies are very highly geared and so, in good years, can pay their interest charges with only a fraction of the profits and retain or distribute much more. (In bad years, the opposite applies.) So it might be expected that after a few years our long-term shareholder of the 'typical' stock mentioned above would reap his reward, and instead of 5 yen dividend a share he would be getting 10, 15, or maybe even 20 yen a share. Not so, for if dividends of 20, 30 or 40 per cent are declared, this is fuel for agitators. So what the Japanese company does instead is to make regular issues of new capital, either free, or at par, or near market price, so that the dividend still tends to remain at 10 yen or less. (On the other hand, if at all possible, they will never cut dividends.) This means that the long term shareholder who begins with perhaps 1,000 shares, for which he paid ¥ 250 each and on which he initially received a dividend of ¥5 per share might after a few years expect to have 2,000 shares at no extra cost on which, the company having done well, the dividend will be ¥10 and the price (because expectations are still high) will be ¥500. It does not take a pocket calculator to work out that this original investment of ¥250,000 is now worth ¥1,000,000, and his before-tax income has risen from ¥5,000 to ¥20,000. Even if the share price and dividend do not go up, he has doubled his money and his return; the moral of the story lies not in the profits, which could be found in most markets with equally good advice, but the presentation which is such as not to alienate labour from capital.

In practice, the investor unless he is a bank, a member of the founding family of the company, or a corporation wishing to secure or symbolise its business relationship, does not hold a share year in, year out. Because of low dealing costs (about 1 per cent each way and no capital gains tax) and because of the trend of the market, he will do much better by switching in and out of shares, taking profits or cutting losses. The reason for this is that whilst individual stocks go up and down quite regularly, often on a cyclical basis, the overall market itself usually moves in a far less pronounced manner.

You will have no difficulty in finding a broker, and several of the brokerage houses have foreigners or Japanese-Americans on their staff to attend to the requirements of foreign clients, whilst even more have English-speaking research and sales personnel. Unless you speak Japanese, you should contact the Head Office of the company to meet such people rather than walk into a street-corner branch where your arrival could create consternation. Basically, any resident foreigner can invest, and arrangements can be made for most non-residents through the medium of an off-shore (Free Yen) account in a Japanese bank. There are certain limitations on foreign holding applied to a very small number of stocks, but otherwise you are free to deal within the Japanese legal and exchange control framework.

You can identify Japanese securities houses interested in securing your business by reading the daily English language press, *The Tokyo Weekender* and *The Japan Economic Journal*, which is the weekly international edition of the *Nihon Keizai Shimbun*, Japan's most respected financial daily. Or you can contact your home broker, or the office of a Japanese broker in your home country before you leave for Japan.

The bible of the Japanese stockmarket is *The Japan Company Handbook*, which comes out twice a year and is the English language version of the *Kaisha Shikiho*.

Foreign interest in the Japanese market is growing, and the Japanese themselves are becoming increasingly involved in foreign markets although 86 per cent of their overseas investment continues to be in the U.S. We shall see more of the Japanese brokerage houses around the world, often playing the same roles as merchant banks — which do not exist as a separate form of economic life in Japan. Radical capitalism is not afraid to let itself be known, for without it Japan would possibly be a socially divided country and would certainly be a less prosperous one — not because of profits or losses made on the stock exchange but because of the special organic structure the Japanese have devised as a part of the group motivating process. And increasingly, our own pensions, insurance policies and savings will have an involvment in what goes on in the Japanese Stock Exchange.

17
Aspects Of Japan's Business Interrelationships

MICHAEL ISHERWOOD

In trying to explain why the Japanese appear so successful we tend to make assumptions based upon the experience of our own society which lead us towards further misunderstanding of the real situation. We take for granted the precepts on which the basis of our business system is organised and conducted. However, in dealing with Japan we have to be much more aware of fundamental procedures and practices that obtain within her business culture if we are to avoid serious handicap.

First, an obvious point: Europeans have spread their culture to many parts of the world — to North and South America, to South Africa and Australasia. What we refer to as the western world. Quite naturally, we tend to overlook the fact that individualism, as we understand it, is a European concept. But throughout most of human history people have lived by the principles of a 'group society'. Indeed, most people in the world today still live within the framework of a group — the extended family, kinship, the clan, the village, the tribe — a society in which the emphasis in daily life is placed upon belonging and responding to the group of people with whom one is intimately involved.

Unlike the western world, however, the Japanese are a group society. In their 250 years of isolation they developed a strong and unique sense of group behaviour. They find it natural to think and act as a group rather than to behave independently as individuals.

So it is important to remind ourselves that the Japanese are not Europeans. By history, culture, language and religion they are Asian. The most significant thing about Japan is that it is the only non-western country to have fully industrialised. The secret of its success lies in the fact that it has taken western technology and absorbed it into its own relatively unchanged social structure. Japan has changed, and changed very rapidly over the last hundred years, but it has managed to avoid the violent social revolution which has characterised the industrialisation of most other countries. Many characteristics of Japanese society today are a clear reflection of the patterns which existed in urban and village life under the Tokugawa administration (1603-1867). The organisation of the feudal village with its strict and formal hierarchy of households can still be seen in the way Japanese companies group together and in the way they organise their internal divisions and departments.

It should come as no surprise to find, therefore, that the Japanese have approached many of the problems of modern industrial organisation in a fundamentally different way from the West. Our historical needs were different as, too, were the means at our disposal. In Europe the development of industry was initially small-scale. It kept pace with technological invention and was slowed by the painful process of trial and experiment. Capital was limited and those who possessed it were encouraged to risk it for the possibility of high reward. As a result, we place emphasis on ownership and there has grown up a division between the interests of the owners of capital and those who invest their labour and skills.

At the end of the last century, Japan needed to develop very rapidly the central institutions of commerce and a core of primary industries in order to avoid domination by the expansionist European powers. It was able to draw on the experience of the West, to import technology and organisational systems as going concerns, and thereby keep experimentation to a minimum. This was done centrally under the initiative of the Government and then when the various enterprises were working successfully they were sold cheaply to whichever private interests were considered capable of continuing to run them. At a time of slender resources, the first requirement was a centralised system of financing and the Meiji planners set up a banking system which was designed to provide finance for industry. Special development banks were formed to provide for the needs of specific industries. The system has survived several periods of reform and remains today the primary source of the working capital needed by industry and commerce. As a result of this dependence on bank borrowing and the fact that even today little reliance is placed upon direct investment, it is normal for companies to have very high debt/equity ratios, usually around 80:20. The system is supported by two strongly entrenched social habits — an unusually high personal savings rate and the tendency for companies to group together in mutual dependence and support.

The average Japanese continues to save something in excess of 20 per cent of his disposable income. It is usual for him to deposit these savings in a bank account and the banks have a very positive and active policy towards their customers encouraging them to invest and to borrow. Banks tend to be the first choice for the investment of savings. There is no highly developed personal insurance market in Japan, a system of Building Societies has not been established (house loans come from banks) and people tend to limit their investment in industry to deposits with security companies. The habit of saving is greatly assisted by the custom that Japanese companies have of paying a bonus twice a year.

There is a variety of reasons why the personal savings rate is so high but it can be summarised in five general categories: (1) The cost of supplementary, non-subsidised, education, such as 'crammer' courses in an education-conscious society where around ninety per cent of children stay on after the official school-leaving age. It is a fiercely competitive system in which the level of schooling attained determines one's future career prospects. (2) A desire to own one's own home which usually means buying the land and building the house on it. Both are extremely expensive in such a densely crowded country. (3) The need to provide for retirement which is compulsory in most big companies around the age of 55. Since life expectancy is now over 70, and in the absence of any significant development of western-style pension schemes, either State or company, the need is very real. (4) Medical care, where again there is no significant development in state assistance and where doctors', dentists' and hospital fees have to be met. Only in the larger companies is there assisted medical and dental treatment for employees and their families. (5) Finally, one can point to the inherited traditional values of Japanese society, frugality and a healthy respect for natural disasters. Both Buddhist and Confucian ethics stress self-denial and a simplicity of lifestyle; peasant communities tend to be cautious and conserve their resources; and the Islands of Japan have throughout history had more than their fair share of disasters — volcanoes, earthquakes, typhoons and wars.

The reliance of industry on borrowed money leads to some other significant differences with western business practice. For instance, the pursuit of corporate profits is not necessarily the primary aim of business. In general, companies compete for growth and market share rather than larger profits. If one is answerable to shareholders, it is necessary to give them a good return on their investment and to compete for high profits in order to attract further investment. It is also necessary to be profitable in order to avoid being 'taken over' by the manoeuvres of rivals in the stock market. But if, as in Japan, the primary dependence is on bank borrowings, then growth is more important than high profit. From the banker's point of view, the loan is for a fixed term of repayment and for a fixed return on the rate of interest charged. High profits make little difference to him (they could be a sign of unnecessary risk) but growth and an increasing share of the market ensure him that the company will repay its debts and will also want to increase its borrowings in the future.

A long-term pattern of stable growth shows that a company is being well-managed and inspires the confidence of the banks who are anxious to off-load the large deposits that are continually being made with them by the private investor. Consequently, companies compete very fiercely for a larger share of the market and keep their profit margins low. Dividends are low by western standards because shareholders are usually the banks and other associated companies who all understand the rules.

★   ★   ★

The fundamental purpose of a company is usually seen to be the fulfilment of its social responsibilities which in practice means that its first duty is to provide and maintain permanent employment for everyone in the company. Responsibility is then extended to society as a whole and to the company's contribution towards the welfare of all people. Hence, the rather pompous statement of principles that most companies adopt and the company anthems that are sung with such gusto by the employees — an expression of loyalty only extended to football clubs in the West. Because companies do not have to protect themselves by issuing large dividends, a major proportion of profits can be ploughed back in reinvestment. The long-term result has been that Japan has been able to reinvest a large part of the national income to stimulate growth — forty per cent of GNP which is roughly double that for western countries.

What is of overriding importance in such a system are the channels through which money is borrowed and this leads to a consideration of two other features of this unique economic system that deserve mention — the connection between companies and the close working partnership between government and industry.

The system of bank financing would not work nearly so effectively if companies acted independently of one another but, because the Japanese tend to think in terms of groups, companies are almost invariably members of some form of grouping. But group does not have the same meaning as it does in the West: it is not based upon the principle of ownership and there is no controlling influence. It might be more accurately described as an alliance of companies bound together over a long period of time in mutual commitment and obligation and based upon mutual trading activities. The whole structure of the Japanese business world is an intricate and complex network of connections and business relationships which defies clear analysis. Most groups are centred around a bank or one of the larger trading companies or around a combination of both.

The trading company will have connections with a thousand or so manufacturing and other companies for whom it acts, not companies that are formally members of the group, but who depend upon the trading company for commercial expertise. The big manufacturing concerns in the group sub-contract most of their component work and so grouped around them will be a host of small manufacturers and suppliers who are dependent for their livelihood on the larger firm. Together with subsidiary operations and joint ventures, the largest company groupings involve several thousand enterprises in varying degrees of association.

The economy is dominated by a number of major groups. The six largest, Mitsubishi, Mitsui, Sumitomo, and those centred around the Fuji Bank, the Daiichi Kangyo Bank and the Sanwa Bank, together command over forty per cent of the total capital employed in Japan and thirty per cent of the total assets. Each of these giant groupings comprises thirty to forty major commercial and industrial concerns. Mitsubishi, the biggest group, includes the largest shipyards, the main aircraft plants, the biggest glass manufacturer (Asahi), the third largest manufacturer of motor vehicles, oil refineries, petrochemical and chemical complexes, the largest brewery company (Kirin), and the producer of one of Japan's most famous cameras (Nikon). On the commercial side, there is Mitsubishi Corporation, at present the largest of the trading companies, the Mitsubishi Bank, various investment and insurance companies, Mitsubishi Estate Company which owns the Maru-nouchi business district in the centre of Tokyo, and NYK which is the largest shipping company in the world.

These groups have no holding company and no controlling stock holder. Each member company is an independent legal entity with an independent management, but they form a loose affiliation with one another in order to accomplish tasks they would be unable to carry out on their own. To help coordinate activities the presidents of the main companies in the group meet regularly, usually at a monthly conference. There is, of course, a continual working inter-change between member companies of the group, particularly with the trading company which takes on the role of organiser and promoter of major projects. (See chapter 19.)

Not all companies, of course, are connected with the largest groups, but the pattern is repeated on a descending scale throughout Japanese business. All companies are connected with others in some kind of similar style of grouping even if it is simply around small provincial banks, because a collective involvement is necessary to safeguard the risks of such a heavy debt-based economy. The system works, and works very effectively.

Because the primary institutions of business and industry were originally organised centrally under government initiative, a close working partnership has been maintained. Government and big business in Japan is not divided in its interests. The education system helps the smooth working of this partnership because those who staff the key ministries and the big companies have graduated from the same leading universities. Key personnel on both sides maintain a close liaison over the years and keep open the channels of communication and introduction. Almost all major business projects are referred to the relevant ministry for consideration and there are daily discussions about trading policy and the strategy that should be adopted for dealing with matters that are in the national interest. In this process, the ministries tend to avoid laying down rigid regulations in the belief that centralised control should remain flexible to suit the changing conditions of any given situation. Japanese refer to this process as 'administrative guidance' and the consensus of opinion that emerges allows for a very high level of agreement about the best direction the economy should take both in the national interest and for the benefit of business. It works because the Japanese are well-schooled in the practice of reaching a consensus and because there is an able and hard-working bureaucracy which is not unduly interfered with by politicians, and there are the large and powerful institutions of the banks, trading companies and principal manufacturers who, through their close and complex connections, exert a strong influence of leadership on the whole business scene.

As in most countries, the government controls the central banking system through the national bank. The Ministry of Finance, through the Bank of Japan, determines minimum lending rates and the amount of deposits other banks have to make. But in a system so heavily committed to bank finance as the principal means of providing investment and working capital, the government exercises an unusually direct and positive control over the flow of funds. It is possible to switch money very rapidly from one sector of the economy to another and to introduce fiscal measures that have immediate effect.

The modern history of Japan has, in fact, seen a continual series of government-inspired changes of direction. Outdated industries have been rapidly run down and replaced with newer technologies which are vital to Japan's continuing success as an economic power. Faced in 1974 with the quadrupling of the cost of her enormous oil bill, the government initiated a new export drive which in less than twelve months earned the increased revenue the country needed. Confronted by the western powers' displeasure over the balance of payments situation, Japan quickly responded with export restrictions, an import policy and a massive overseas investment programme. The effects have not been quite so dramatic as previous changes in direction because of the prolonged economic recession, but they have nevertheless demonstrated the ability of the system to cope quickly with these events. It is made possible by the ability of the government with the close agreement and support of industry to make funds easily available to any sector which conforms to the consensus of opinion about national policy.

It is not easy for others to learn lessons from the Japanese experience because none of these fundamental elements or inter-relationships would work out of context. The banking system which depends on a high personal savings rate, the trading companies with their enormous size and diversity of interests, the collective nature of the Japanese which gives rise to the groupings of companies in a most complex way, are all indispensable elements. But if we are to profit by working with them either by developing markets for our goods in Japan or cooperating in major project work in other parts of the world, it is essential that we are aware of the way they differ from us in their approach to economic organisation and that we appreciate the pressures and problems that influence their thinking.

18
Management Style

ROBERT J. BALLON

Adversity is the acid test for any human endeavour but prosperity may well be the test to prepare for such adversity. It was no mean achievement for Japanese managers to lead their companies through the neck-breaking speed of economic growth through the 1960s. Then came the 1970s with the floating of the yen, oil crisis and a bout of two-digit inflation, drastic upvaluation of the currency, massive surplus on the balance of trade and international trade problems. Japanese management faced an enormous challenge and . . . survived on the strength built up during the prosperous 1960s. The impressive characteristic of Japanese management is continuity.

Japan's history and culture, not to speak of her economic success, have caused Japanese managers to develop a style of management which is their own and is, generally, very successful — at least by Japanese standards, and indeed by some of our own. Corporate debt, for example, government regulations and personnel adminis-tration are used in a way that might be rightfully questioned outside Japan. However, they are part of what the Japanese challenge is all about.

The interpretation of the past is often coloured by current circumstances. In pre-war Japan, it was customary among scholars to consider business managers as the descendants of the *samurai*, who for centuries had dominated society as the nation's élite. After the Second World War, to be more in tune with modern trends, managers came to be described as the descendants of the merchants who, in feudal times, made it possible for society, including the *samurai*, to survive. It would probably be nearer the truth to describe the pre-war and post-war business executives alike as the descendants both of the *samurai* and the merchants.

Before the Meiji Restoration of 1868 the acquisition of wealth, though officially frowned upon, was the real goal of both the higher social class, the *samurai*, and the lowest class, the merchants. But it was rather as the head of the extended family acting on behalf of his own personal family that an individual desired wealth. The concept of property was depersonalised, in the sense that property belonged to the extended family personified in the family head.

The acquisition of wealth, though a prerequisite for Japan's industrialisation, was not its cause, since the first faltering steps towards industrialisation were taken by the political leaders of the Meiji Restoration rather than by any future captains of industry. By origin, therefore, Japanese industrialisation is different from that of the West. It was only after more than a decade of hesitation, in the early 1880s, that the private sector finally awoke to the new opportunities; surprisingly, too, the government was happy to turn over to the budding industrialists most of its rather dismal attempts at industrial development.

The West was to be the model as far as industrial technology was concerned. However, industrialisation was not the end but merely the means by which Japan would be able to take her place in the world. Thus, there was no question of modifying Japan's spiritual values in the industrialisation processes. The dichotomy was tersely expressed in the slogan, *wakon yosai*, Japanese spirit and western skills. To this day the slogan remains at the root of Japan's success, if not also at the root of the international misgivings about her success.

The economic success of post-war Japan remains a puzzle as long as it is analysed according to western norms. The key is not to be found in identifying so-called favourable circumstances which, after all, were at the disposal of other industrial nations, but rather in something that could be considered and described as typically Japanese — something which allowed Japan to capitalise on these favourable circumstances. This 'something' is particularly evident in the structure of the Japanese enterprise which can be examined from three basic viewpoints: its employment system, its capitalisation and, finally, its competitive position.

Employment system

In industrial terms a Japanese does not identify himself by his profession, occupation or skill, as a westerner would. He identifies himself and is identified by society at large, according to his work-place (*shokuba*), i.e. the company that employs him. In the context of industrial society what the westerner sees as the role of a profession or occupation and the expectations he attaches to it come to the Japanese from the enterprise. It is from the enterprise that the Japanese derives both security and satisfaction. Employment, therefore, is not regulated by a contract that hires a skill; it is a life relationship that attaches a man to a specific organisation. Labour contracts are indeed the exception in Japan.

In the context of the work-place, there is also no fundamental distinction between management and labour. The company employs managers and managed alike. The image of a coin comes to mind: one side of the coin is management, the other is labour; they are definitely different, but they converge to make the coin, the company. In that it assumes not a divergence but rather the convergence of different interests, the style of management in a Japanese enterprise will not be readily comparable to, nor imitable in, the West.

A striking illustration of this unique Japanese outlook on the enterprise is given by the type of labour organisation. Japan has no 'labour union' in the western sense: she has a peculiar form of labour organisation called an 'enterprise union'. This union restricts its membership to all regular employees of the enterprise and, as a rule, excludes outsiders to the company from its bargaining with management. When an employee quits, his union membership is automatically dropped, and if the company were to go bankrupt it would also mean the end of that particular labour union. It does not mean, however, that 'enterprise unions' are mere 'company unions', in the sense that they could be or are dominated by management. Conflict there is, but it tends to be resolved within the context of the company rather than outside the company or on ideological grounds.

Not less striking is the fact that, in general, the same salary system applies to ordinary employees and to operating managers, including department heads. The result is that at the bargaining table the managers involved are, in fact, negotiating their own remuneration. These negotiations usually take place at a pre-set time, the so-called annual 'Spring Wage Offensive'. In post-war Japan, the spring is heralded not only by the cherry blossoms, but also by wage struggles. As a result of this annual recurrence over the last 25 years, wage levels are raised simultaneously throughout industry — thereby having little impact on the competitive relationship among firms: and, of course, the production schedule is readily adjusted to allow for some spring problems.

All this goes a long way towards explaining why the Japanese, in western eyes at least, appear to be 'hard-working' people. That may be so. On the other hand, you would be right in assuming that two entirely different work ethics are involved here. Not even a Japanese enjoys work for the sake of work; the difference in output, as it were, results from the fact that work is considered less as a punishment or a form of exploitation and more as the normal way of acting. It could probably be said that a Japanese does not work for a living; he considers work a way of life.

Capitalisation

It is well known that whereas western companies need something like 40/60 debt-equity ratio, Japanese companies, since the end of World War II, thrive on a 80/20 ratio. Notable exceptions there are, but they appear to confirm the rule.

In sociological terms, the 60 per cent equity stands for western corporate individualism, i.e. the independence of the individual corporation. The way to strengthen this equity is by profit maximisation, as a better price-earnings ratio promises increased access to the securities market. On the other hand the 80 per cent debt in Japan should not be interpreted as control by the banks; it stands for the interdependence of all corporations.

The debt position of Japanese corporations, notwithstanding the Commercial Code, explains why Japanese top management does not feel really accountable to shareholders, who could give it its independence of action or . . . dismiss it altogether. Top management, especially the company president, plays it safer by involving itself with the application of interdependence, as expressed by the predominance of industrial groups, the major role of trade associations, the partnership with government agencies and, last but not least, by the debt position. It is thus not the banks that 'control' debt; banks are but part and parcel of the system.

Although the debt-equity ratio in pre-war Japan was similar to that current in the West, the present 80 per cent (which to a westerner would mean courting bankruptcy or takeover by the banks) has meant rapid expansion in the post-war years. Progressive companies had to double their operations every two or three years. In most cases this would hardly have been possible out of retained earnings; it required fresh capital in abundance. This form of massive capitalisation was not sought on the stock market as long as custom required that new shares be issued at par value. On the other hand, besides the fact that loans are acceptable as a deductible business expense, the banks by their loans and the discount of promissory notes, backed up by the Bank of Japan, were most anxious to contribute their funds to industry. These funds in turn were largely provided by individual depositors (the general public) whose saving rate is, today, over twenty per cent of disposable income.

Corporate management is not particularly interested, therefore, in maximising profits that improve the price-earnings ratio. It is much more concerned about minimising the cost of borrowed capital by expanding its market share. This concern for the corporate market share is not reserved to managers; the work-force itself is extremely keen in its regard for this objective as it defines the company's industrial and social ranking.

Competition

It is thus to be expected that the relationship between Japanese companies is extremely ambivalent — at once competitive and cooperative. It should be remembered that there is really no Japanese equivalent for the western word 'competition'. Last century when Yukichi Fukuzawa, scholar and school administrator, largely responsible for the introduction of business administration into Japan, was translating an English textbook on economics, he ran into the word 'competition'. He rendered its meaning by fabricating a new word, *kyoso* (*kyo* meaning race, and *so* fight);he was criticised not so much for his semantics, but for introducing an 'un-Japanese' value!

Even today the term is not fully accepted. When Japanese businessmen use the term *kyoso* they always prefix it with *kato,* which means 'excessive' competition. The implication appears to be that competition, which in the West stands for the normal relationship between enterprises, in Japan is simply considered as an excess. This 'excess' derives precisely from the dynamics of growth and the drive for market share: it leads corporations into expensive investments in facilities, not justifiable today, but required for tomorrow. The same 'excess' is at work among the sub-groups that compose any large group, be it the company, industry generally or the economy at large. It is further compounded by the social technique of 'crisis', so prevalent in Japanese society. Again let us compare this situation with the human family. For most human beings it is a daily experience to live in a family; but an explicit consciousness of family life is revealed at a time of crisis. This is exactly what happens in any Japanese community, be it social or industrial. If an emergency can be construed or is actually on hand, the crisis atmosphere calls for the mobilisation of all energies within the organisation. Sacrifices are then accepted without question, and petty rivalries are overlooked; the emotional commitment reaches its peak.

All this has little to do with the western concept of competition, but in the Japanese organisation context it stands for *kato kyoso,* excessive competition. It is an active ingredient of the dynamism of corporate growth as well as of national economic growth. This brings us to another aspect of Japanese-style competition, one that — to put it mildly — leaves the non-Japanese businessman almost permanently bamboozled; the partnership between government (i.e. bureaucracy) and business. This partnership is completely remote from the classic planned economy.

The crux of the matter is a question of continuous interaction between government and business, for which the foundation is not the written text of the law, but the imperative of Japanese social dynamics. This pattern has been set by tradition, namely the tradition of the warrior-bureaucrat of the Tokugawa Era (1603-1868) who made, interpreted and administered the 'law' ('law' to be understood not as a written text, but as a rule of equity). The poles of this continuous interaction are administrative guidance and consensus, both effect and cause of the 'equity' role of bureaucracy.

Administrative guidance *(gyosei shido)* as provided by the Japanese bureaucrats to industry has no legal basis, and therefore no legal sanction. The penalty for not abiding is not direct but indirect, precisely through the mechanism of continuous interaction. It is essentially a case-by-case problem-solving process. The matter could probably be best understood by comparing it to traffic regulations. That traffic runs either on the left or right side of the road is an apparent universal principle but it makes sense only if and when there is traffic, namely if and when two cars drive in opposite directions...

Administrative guidance is not so much the actual application of some universal principle, it is rather the manifestation of a direct initiative taken by the bureaucracy, sometimes arbitrarily but mostly controlled by consensus.

Consensus in Japanese society does not stand for acquiescence or unanimity nor does it stand for an inherent unity of views in regard to broad objectives. Consensus is a style of life, not a mental process. When consensus is reached it does not mean that all parties involved speak with one voice, but that a given issue (as was the case for the previous issue, and will be the case for the following issue) is under consideration by all concerned: it means that execution is taking place. The consensus is formally achieved in joint consultative committees *(shingikai)* and in trade and industrial associations.

For example, MITI (Ministry of International Trade and Industry) regularly reviews the annual investment plans of such capital-intensive industries as steel and synthetic fibres. The matter has been discussed in the industrial association, and is discussed again in the *shingikai,* where the key members of the association meet with the appropriate government officials and private experts. MITI contributes its own views and acts as conciliator and industry responds in view of its own needs and the ability of the ministry to service these needs. More rarely the conciliator turns into an arbitrator and establishes a rule of conduct that will be enforced, such as when cartels or the minimum capacity of any new plant are involved. In all cases bureaucracy remains an active participant in the consensus process. But it would be a gross exaggeration to say that bureaucrats 'run' the economy. Industry, for its part, is also outspoken about its own views. Economic organisations, in particular *Keidanren* (Japan Federation of Economic Organisations) and *Keizai Doyukai* (Japan Committee for Economic Development), repeatedly make it clear that the formulation of national policies is not an exclusive domain of officialdom. Business policies are, by and large, kept outside the political arena, certainly more so than in the western industrial countries. It was all admirably stated by Lockwood:

> The metaphor that comes to mind is a typical Japanese web of influences and pressures interweaving through government and business, rather than a streamlined pyramid of authoritarian control. Perhaps it is just as well. Business is somewhat shielded from government dictation by inter-agency and inter-group tensions. Its own disagreements in turn tend to diffuse its counter-influence in politics. The danger, of course, is a tendency to indecision and drift where national interest may call for clear-cut decisions. Only the biggest decisions go up to the Cabinet. There they may still encounter interfactional and inter-Ministerial rivalry, as in the annual contest over the budget. Even then opportunities for noncompliance down the line are considerable. A web it may be, but a web with no spider. What makes the system as workable as it is, no doubt, is a strong *esprit de corps* in the higher ranks of the civil service, and a common social background and university training among leaders in both government and industry.*

*William W. Lockwood ed. *The State and Economic Enterprise in Japan* (Princeton University Press, 1965) p. 503

The Japanese manager

Against this background, let us now turn to the Japanese manager in his office.

Post-war Japan has been characterised by the enthusiasm of

Japanese management for importing 'modern' technology of production as well as management. This was done mainly from the United States by numerous management missions, a flood of literature, translated and original, and an equally powerful flood of seminars and international conferences, Throughout, the concern was to assimilate, and assimilation has taken place to the extent that today there is a 'modern' (though not necessarily 'western') style of management in Japan.

In the process, Japanese management has become 'professional'. This qualification, though ringing familiar to western ears, should be understood in the Japanese way. Contrary to pre-war practice, when managers were promoted on the basis of their kinship to a given family or partisanship to a political régime, in post-war years management positions have been obtained on the basis of alleged or demonstrated competence. Such competence, however, is not evaluated according to universal 'professional' standards applicable to any individual anywhere, but according to the needs of a given corporation primarily in regard to those who come up from within its own ranks. In genaral terms it could be said that what is 'professional' about post-war Japanese management is simply the fact that it uses management techniques: there is no 'profession' (at least not yet) in the sense of supplying an individual with credentials which could be meaningful in moving from one company to another. As stated earlier, the industrial identification of a man, be he an employee or a manager, is derived from his workplace.

There are two common forms of mobility for managers: *amakudari* and *shukko*. *Amakudari* (descending from heaven) is a colloquial expression used when high-ranking government officials, upon (voluntary) retirement from the civil service, move into a directorship position in some large corporation in the industrial line they were active in previously. Thus, some directors of financial institutions come from the Ministry of Finance; in the pharmaceutical industry from the Ministry of Health and Welfare; in the construction industry from the Ministry of Construction. In the private sector former officials do not usually become company presidents, but the Board of Directors is their safe haven and their presence there is the lubricant of effective government-business partnership. The top management of public corporations and other government-related enterprises is largely composed of these former officials.

Shukko (transfer) is a very common practice whereby a parent company sends some of its executives to a related company for a limited period of say two or three years. This is the way most joint ventures are staffed with managers from the Japanese parent. The 'transfer' can also be permanent, as in the case of an executive of the parent firm who, upon reaching the mandatory age limit, is promoted to a higher position in a related firm. The purposes behind such a move are varied, but always very specific; there are four possible reasons.

- To unload retiring officers, who are not good enough to be kept or would interfere with the promotion scheme at the parent company.
- To control the related firm from within, as when a bank sends one of its retiring officers to a major borrower, or when the decision is made to 'rescue' a failing company.
- To strengthen ties with a particular industrial grouping, like Mitsubishi or Mitsui, or a smaller more recent group centred upon the personality of its founder, as is found especially in the industrial cluster around a private railway.
- To train younger executives by broadening their business experience.

Outside these two cases of mobility, the rule is that managers are promoted from within the company. But what happens when a company is expanding very rapidly from a rather narrow starting point? 'Head-hunting' in Japan comes mainly in two forms.

Recruiting in the market. It is always possible that an experienced manager wants to quit for honourable reasons. This is rather extraordinary in Japanese companies, but fairly common in foreign ones. Banks, in particular, sponsor their own placement offices which look for jobs for their retiring employees and managers. In recent years, several private executive-recruiting agencies have been started; they are , however, mostly active for foreign enterprises and 'hunt' mainly from foreign enterprises. The public Employment Security Offices sponsor so-called Talent Banks specialising in the placement of older and middle-aged employees, including people who exercised some managerial responsibility in smaller firms.

Recruiting through acquaintances. This is usually a more successful route than the former; it presents the invaluable advantage (in Japan) of involving a sponsor. For example: an excellent manager in a large corporation, reaching the age of fifty-two or three, cannot be promoted for lack of vacancy on the Board of Directors: his president will be on the look-out for a good assignment for his former subordinate. There is also a certain unspoken cooperation among Japanese companies that 'tolerate' scouting by up-and-coming companies in dire need of experienced managers, thereby relieving some of the pressure from below within the older established company.

Such a lack of mobility on the labour market is a major handicap to foreign operations in Japan. The problem is easing only very slowly; it constitutes a formidable non-tariff barrier that will protect

Japanese industry for several more years to come against the establishment of large hundred per cent-owned foreign establishments as well as takeovers.

Top management

Turning now to the management hierarchy, the customary division in the West is: top, middle and lower. These terms are used extensively in Japan, but they do not appear to fit the western structure too exactly. There is a top-management level, where the key figure is the company president: below this level there is the operative management.

The Board of Directors' power is normally concentrated in two or three of the senior directors, who probably are also representative directors; they necessarily include the president who is the chief executive officer of the company. Heading the Board is the Chairman, a title often acquired by the president or given to the out-going president, thus making the role more of an honour than a responsibility. A post-war peculiarity is the fact that several directors are usually full-time operating managers. (Senior managing directors often include division heads, and among the other directors are found managers of major plants or even presidents of major related companies.) With so much of the day-to-day business management represented on the Board, its major function of policy-making is heavily concentrated in the hands of the chief executive officer, the president. In normal circumstances, he may expect the rather passive acquiescence of the Board.

Most corporations have another top-management organ that is not required by law: it is generally called *jomukai* (Executive Committee) or some similar name. If there is any policy problem, it will be discussed in this committee by the same people who meet in the Board, with the exception usually of the Chairman of the Board and a few outside directors. But in the presence of so many operating managers, the Committee will be concerned with routine matters.

Generally speaking, therefore, Japanese corporations at the top are weak in their policy formulation. But this weakness is probably less a matter of management than a feature of the Japanese character that is better at reaction than at action.

The company president

It is thus the president, as chief executive officer, who is expected to provide the kind of leadership that the Japanese industrial organisation needs in order to survive both domestic and international competition. In fact the president monopolises much of this leadership. But leadership in Japanese terms has two basic qualities:

- It is not imposed from the outside: it answers an internal need.
- The qualification as 'leader' is less a matter of personal talent or merit than acceptance by the organisation. It supposes that the 'leader' shares in the mutual emotional commitment within the group. The leader-group relationship, therefore, lasts beyond the time and the circumstances of any given task.

Not unnaturally, therefore, within the corporation the senior directors usually form an intimate group surrounding the president, while they themselves command similar retinues among junior directors and other managers. As a result, any challenge to the incumbent presidency will come primarily from within the organisation. On the one hand the president must hold together his staff and work-force so that the enterprise stays ahead by increasing its market share. This is his internal function.

It is this internal function that is at the root of the sweeping generalisation that Japanese management is paternalistic, if not authoritarian; Japanese scholars, often of Marxist leaning, brand it as feudalistic. On the other hand the president must also react flexibly enough to the demands on the organisation from lateral commitments: the banks, the industrial group, the trade association, the government. This is his external function.

The external function of the company president is much more apparent. Apart from the fact that multiple presidentships are fairly common, a large proportion of the president's time is spent pursuing public relations activities. He has to meet his counterparts in the industrial group and in the trade and industrial associations:he has to cultivate personal contacts with financial backers, with major clients and, last but not least, with government officials.

The role of the company president, therefore, is essentially a 'social' one from both the internal and external point of view, rather than a direct involvement in the company's real business matters. His leadership rests more on the qualities revolving around human experience and contacts than on technical qualifications. Not surprisingly, Japanese company presidents are usually in their late 60s.

Younger company presidents in their late 40s or 50s are more likely to strive for personal responsibiliy in the company's operations. In this, they are usually helped by a more direct contact with overseas operations and the fact that bribery cases in the late 70s cast a shadow on the top management of some large enterprises.

Operative management

Operative management is essentially exercised at two levels; the *bu* (department) and the *ka* (section), a subdivision of the *bu*. Truly functional titles are therefore *bucho* (department head) and *kacho* (section chief). Such is the theory. The practice, however, has all the complexity of an organisation continuum which becomes more marked as the enterprise gets larger. Managers and 'managed' form one and the same reality, the corporate collectivity. The same salary system applies to supervisors, clerical employees and manual workers; holidays and pension rights are the same; dress is the same and there is no discrimination in the availability of canteens, toilets or car parks.

As a rule, the Japanese organisation is a line organisation, though the 'line' followed is not exactly the line of authority; it appears to follow rather the 'flow of work' concept. Authority, and therefore responsibility, is not distributed and allocated among discrete individuals at discrete levels; it is diffused throughout the organisation. Japanese scholars and managers often contrast the 'top-down' approach of western management to the 'bottom-up' approach of the Japanese. The difference goes back to the independence of the individual *versus* the interdependence of the member of the group.

In this context, many managerial titles are no more than a mere recognition of status, with little if any definition of the function and responsibility of the positions involved.

It could be said, therefore, that Japanese companies are beehives of 'human' activity rather than 'business' activity — in the western sense. What is more, it is often said that the Japanese are hard working. This is an exaggeration; it would be more correct to say that they keep very busy — busy, that is, with the business of human relations within the frame of the organisation, rather than busy performing physical or mental work. It is little wonder that company presidents become so deeply involved in personal matters, having to act as a kind of umpire over the lively factionalism in their organisation.

Slow decision

In a Japanese organisation, with its heavy reliance on human considerations, time is required to reach what appears to be a decision: in fact what is arrived at is a consensus for execution, as noted earlier. Management is keenly concerned with involving in the decision-making process everybody who will be involved in the execution. This is achieved through repeated formal discussions and formal meetings, which are often aided by the circulation of a document (*ringisho*) stating the matter on hand. Thus, when a company is considering the construction of, say, a new plant, nothing much can be expected from a mere decision — that in itself is meaningless unless, or until, everybody has been involved. When finally the decision is taken, it is tantamount to the start of work operations.

Slow execution

The western decision-making process largely revolves around the responsibility of the decision-maker. Once the individual who will take this responsibility has been located (he is selected almost automatically by his position on the organisation chart) alternative strategies are considered, and what is thought to be the best one is selected. Now starts an often lengthy process leading to execution, as the various levels of the hierarchy are informed about the decision and requested to cooperate and prepare to put it into effect.

Following this schematic contrast, the difference between the Japanese and the western process is not so much a matter of time. Whatever amount of time is required by the Japanese *before* the 'decision' is made, a similar amount is required by westerners *after* the decision is taken. The basic difference between the two approaches is to be found in the style of execution. It is obvious that following a decision by consensus (Japanese-style) the execution of that decision by those involved will be highly motivated in the sense that 'this is *our* decision'. On the other hand, the execution of a decision 'on command' remains just that — motivation being a very secondary implicit element.

However, in view of the factionalism in Japanese companies, it would be quite wrong to see Japanese consensus as a kind of passive acquiescence, or as something that springs forth spontaneously. It is not even decentralised decision-making. The group and departmental efforts are duplicated; authority mostly overlaps; reporting procedures are often confusing; communications end in bottlenecks and specific assignments are often too narrow to be truly challenging, and so on. All these weaknesses result from lack of western-style 'functionalism'; but such or similar weaknesses are also found where the concept of functionalism supposedly prevails.

The single most important reason why Japanese companies do not stop dead in their tracks is because they have operating managers. Much more than top management *per se* (let us remember that many directors are also managers), they are accountable for the performance of the human organisation of which they are a part. But accountable to whom? Essentially, they are accountable to the organisation, since management is a function of the *shokuba* (the place of work). Managers, of course, pursue their own career, but it remains understood that this career is with the company.

Business Career

In the broadest terms, there are five fundamental steps in the career of a Japanese manager:

1. The managerial career begins with college (four years), followed by graduation at age 22, followed by selective recruiting. The implications of the school clique are an important consideration for the simple reason that human dynamics work better among people sharing a common educational background. It has to be said, however, that the clique system is not quite as strong as it was. Nepotism has little impact and is generally frowned upon. There is also the solidarity, for many years to come, of all those who joined the company in the same year, a peer group, we may say. These internal factions contribute powerfully to corporate dynamism, though not always in a beneficial way. But, in adverse times, when sacrifices are demanded, they offer a welcome moral support.

2. For the next five to ten years the managerial candidate will work as a general office clerk, rotating jobs in the various sections and plants of the company. In each instance, the technical demands upon him will be great; they are to be answered primarily by on-the-job development and initiative. (These are the young people who surround the senior men who negotiate with foreign business representatives.) Characteristically, all of them will be union members, and to become a union officer is often the sign that one is ripe for being promoted to the first step of the managerial ladder. At present one out of six board members is a former union officer.

3. After ten or more years of employment in the same company the first threshold is reached, namely promotion to the position of *kacho* (section chief). This is as far as some will ever go, but the practical experience they accumulate as years go by provides the company with essential expertise. For others, it now becomes clear that they are on the way up.

4. For these especially promising executives, the real need now is to widen their 'experience' and not, as might be assumed, to widen their technical qualifications (though they are regular attendants at seminars and other training activities). As numerous surveys have uncovered 'experience' is, in the eyes of their superiors, the major qualification for further promotion. Experience is seen as something that cannot be taught; it is a function of age, and is manifested in what could be generally called human relations. In this context, the manager is expected to develop into a generalist rather than a specialist.

5. After a total of some twenty years of employment another threshold is reached, that of *bucho* (department head). Because of the shortage of vacancies at the top, or because of individual shortcomings, some managers will stay in this position until retiring age, between 55 and 60, at which time they will then be transferred to a related firm or the like. Others, round about the age of fifty-two or three, will be promoted to a directorship, while maintaining their *bucho* position; to them the normal retiring age limit does not apply.

In any consideration of the career of a Japanese manager, however, what must be kept in mind is the fact that group solidarity and factionalism exert strong pressure from below. As with the company president, managers at all levels have to display leadership. They are not supposed to do the work themselves, but they must have it done by their subordinates while at the same time they themselves remain involved in it, which can be difficult.

This system has thrown at Japanese managers what they consider their greatest challenge, the so-called generation gap. They all complain that today's younger generation is rebellious (forgetting that they themselves years back chaffed under their superiors). As with management systems throughout the world the best cure for subordinates' dissatisfaction is promotion. However, in the context of what was so far called lifelong employment and now pointedly begins to be called stable employment, developments in the late 1970s are forcing a re-appraisal of the promotion system. The slower rate of growth and the rapid ageing of the working population are severely limiting promotion opportunities, functional promotion as well as status promotion. Many companies go through an agonising review of their operative management structure, if not also on anybody above 45 carrying a title. Often the best proof of one's worth is the following: whether he knows exactly what is involved or not, a manager must promote initiative from below by sponsoring the latest managerial technique available, before any other section in the company does!

It is now a hundred years since Japan started to adopt and assimilate aspects of western business management techniques. They were adopted with as much eagerness as was the introduction of western technology. In pre-war as well as in post-war Japan, whatever success has taken place could not have been possible without the West's contributions. However, the fundamental question is this: Why has Japan been successful? The answer must surely lie in the human dynamics at work in the industrial enterprise and the economy at large. It would be too western an evaluation to give all the credit for this success to the managers: the credit has to be shared among all the Japanese. Nevertheless, it must be said that

Japanese managers have displayed an extraordinary flexibility in leading Japan from an impecunious economy to one of great affluence. So much so that today managers believe they have been singled out as scapegoats for the cost of this affluence — namely pollution. Now a new stage of management development is in the making. On the domestic scene, if the last one hundred years are in any way typical, it appears that management philosophy and practice will continue to change rapidly, without necessarily becoming 'more western' or 'less Japanese'.

However, of far more consequence for the future evolution of Japanese management is the fact that, as Japan increases her foreign direct investments, more and more Japanese managers will become increasingly involved in overseas operations. It is only very recently that Japanese corporations realised (with genuine dismay) that their overseas operations are overwhelmingly staffed by Japanese — an understandable situation so long as *international* operations meant *Japanese* operations overseas. A new and brave 'party line' however is now in the process of formulation, namely management participation rather than management control. It may well be that, after some more traumatic experiences, this approach will be acceptable to both the foreign and Japanese parties. Even so the Japanese side will probably retain the advantage, since its domestic style of management is a question of participation rather than control.

19
Japanese Trading Companies In Transition

SADAO OBA

Since the oil crisis of 1973 big Japanese trading companies have been facing major problems. The crisis has been so bad that Ataka & Co., the tenth largest trading company, went bankrupt and was forced to merge with C. Itoh, one of the top five.

In the hyper-boom years just before the oil crisis, people enviously said, '*Sogo Shosha* is the Nation' or 'It is like an unsinkable battleship'. Then, suddenly, it all changed. The oil crisis and the following recession saw to that. This time it was 'The trading company is multiple recession industry' — at least according to the president of one leading trading company. The point was that steel, petrochemicals, fertilisers, sugar, shipping, and so on — all typical trading company interests — were also the typical industries severely damaged by the recession, as well as by the competition coming from newly industrialised countries such as Korea, Taiwan and Mexico. What is more, during the recession, trading companies were obliged to assist certain companies they had dealings with by giving more credit or cheaper interest rates to prevent them going bankrupt. This rescue work, in addition to the general decline in their own export and import business, put trading companies in a very difficult situation.

After several difficult years, the trading companies are once again regaining their powerful position in the Japanese economy and are again active in business. In the business forecast for FY 1979 (April '79 to March '80) the total turnover of the top nine trading companies will be 8.8 per cent more in volume than in the previous year. The increase comes mostly from the higher turnover in basic energy supplies (petroleum, coal, gas and uranium). For example, in 1975 trading companies imported 17.5 per cent of all crude oil imported into Japan, but they increased their share to 26.4 per cent in FY 1978. The import of crude oil into Japan had been the virtual oligopoly of major oil companies plus a few Japanese petroleum refineries, but the share of the trading companies in crude oil trade has been constantly increasing and now no one can ignore the role of trading companies in securing essential energy supplies for Japan. Here, it is interesting to note that after the Iranian Revolution in

March 1979, it was six *Sogo Shosha* and two Japanese refineries that made the first contract for 600,000 barrels per day of crude oil with the Iranian National Oil Corporation.

The rise of the *Sogo Shosha*

The trading companies' power comes from their flexibility and vitality. There are several factors which have contributed to this powerful position.

First, they have taken positive and vigorous advantage of the rapid expansion of world trade and the internationalisation of Japan's economy. Nine *Sogo Shosha* handle about 50 per cent of Japan's total exports and about 60 per cent of total imports. (Total turnover of these nine in FY 1977 was ¥47,008 billion — U.S. $211.75 billion, which is similar to the GNP of Britain in 1977.)

Secondly, they have brought their 'organisation function' into full play. They organise their resources, man-power, money, commercial and technical know-how for originating very big projects both in Japan and overseas. They also play an organisational role in establishing consortia for newly-emerging industries, such as oceanology, space, urban developments, leisure and research. To develop these industries, the general trading companies act as a go-between, linking the manufacturers of different projects with the banks, insurance companies, transportation companies, etc. within their group and also with companies belonging to other groups. They gather information from all over the world and furnish key managers for newly-established specialised companies.

This organisation function is most significant in overseas projects and investments. For instance, Mitsui is the largest Japanese overseas investor with accumulative overseas investments and loans of ¥2046.8 billion (U.S. $1100 million) at the end of March 1978. Mitsubishi had ¥1443 billion (U.S. $650 million). Other *Sogo Shosha* also have substantial overseas investments and loans. Because of this enormous sum of foreign investments and world-wide trading networks, Japanese trading companies have acquired a very significant and influential role, not only in Japan, but also in the Far East and South-East Asia generally.

Foreign companies wishing to participate in these fast-growing industries in Japan can do so with the help of the general trading companies as recent developments show. For example, Holiday Inns Far East Ltd., (a subsidiary of the biggest hotel chain in the world) has been cooperatiing with C. Itoh in establishing hotel chains in Japan where they have so far opened four hotels. In 1977, Mitsui & Co. established a joint venture with British Leyland, 'Leyland Japan', to market the Jaguar, Rover and TR7 in Japan.

Since then, the joint venture company has been selling more and more Leyland cars in Japan. In 1970, Heublin (U.S.A.) established a joint venture, Japan Kentucky Fried Chicken, in collaboration with Mitsubishi Corp. and today has 196 shops throughout Japan.

Thirdly, the sphere of activities of general trading companies is not limited to Japan but covers the whole world; they export U.S. grain to Europe, British machinery to South-East Asia and German chemicals to Latin America. They are also actively engaged in marketing products and commodities; in exploitation of natural resources (iron ore, copper, coking coal, petroleum, timber, etc.); in farming, fishing and manufacturing, and in developing infrastructures.

Their overseas investments naturally contribute to economic development in the host countries. These investments also contribute to the expansion of third country trade. For example, Mitsui is going to sell huge quantities of petrochemical products to be produced by Iran-Japan Petrochemicals Corporation, Bandal Khomeini of Iran, of which the total investment is about U.S. $3 billion. Mitsubishi will launch a similar petrochemical complex in Saudi Arabia and will be marketing the products in the middle of the 1980s. Such huge investments, although they are risky, enable the *Sogo Shosha* to secure a new source of products for third country trade in place of Japanese products which are gradually losing international competitiveness.

Third country trade is also promoted by acquiring existing distribution networks. In 1978, Mitsui purchased grain elevators in the U.S.A. from Cook Industries and is shipping grain and corn world-wide. Mitsui is now regarded as one of the biggest grain merchants in the world. Mitsubishi have followed Mitsui and in the summer of 1979 decided to purchase the majority share of Copel Inc. (U.S.A.), and to increase their grain shipments to South-East Asia.

Third country trade has been especially encouraged by the top managements of trading companies; in fiscal year 1977 the total third country trade of the top nine *Sogo Shosha* amounted to ¥4,039 billion (U.S. $18.2 billion) and 9.1 per cent of the total turnover. Although third country trade is usually more risky than the ordinary import-export business, it will be gradually expanded to 10 per cent of trading company turnover in the very near future.

As a result, some foreign governments, which have been watching the remarkable activities of Japanese trading companies with great interest, intend to develop similar general trading companies in their own countries in order to increase exports.

Fourthly, the immense financing power of the general trading companies has expanded rapidly. In this world of economic growth, money follows if a project is feasible. General trading companies have originated many new methods of securing bank loans and raising the urgently needed funds for such necessary objectives as the economic rehabilitation of Japan, export financing and economic development, the exploration of natural resources and the construction of infrastructures in overseas countries.

Trading companies also provide credit for manufacturers, wholesalers or mass-retailers. Although the requirements from 'grown-up' companies are not so great as in the hyper-boom days, trading companies are still taking care of many manufacturers — in some cases, charging lower interest than the banks. This financing power is reinforced by the highest reputation of the trading companies, not only in Japan, but also in the international financial market.

There are many specialised marketing companies under the wing of the trading companies. In most cases, the trading companies keep the controlling shares and thereby control the operations and development of these companies.

Fifthly, the general trading companies have established throughout the world a most up-to-date and well-organised communications network. Mitsui, for example, was the first company to introduce a world-wide private telegraph circuit which interlinks 50 offices in Japan and 130 overseas offices in 77 countries. Through this private telegraph circuit, 25,000 messages are sent every day by utilising O.C.R. (Optical Character Reader). Mitsui pays ¥2400 million (U.S. $10.8 million) annually to Japanese and overseas post offices. Besides telex fees, Mitsui has an annual telephone bill of ¥2000 million (U.S. $9 million) and a postal bill of ¥600 million (U.S. $2.7 million). The total communications expenditure annually is ¥5000 million (U.S. $22.5 million). Other trading companies pay similar enormous amounts for their communications systems.

Because of these world-wide communication networks, *Sogo Shosha* can respond immediately to very sudden international political and economic developments, such as monetary realignment or an oil crisis.

As international operators, a facility with languages is an important part of the trading company communications network. Most companies, therefore, encourage their staff to learn foreign languages, pay for their tuition and send them to foreign universities.

Multi-nationalisation of the *Sogo Shosha*

Recently, the *Sogo Shosha* have put their first priorities in multi-nationalising their organisations. Several *Sogo Shosha* are locating senior managing directors in regional headquarters such as New York (covering North and Latin America), London or

Brussels (covering West and East Europe and/or Africa) and Sydney (covering Australia and other Pacific islands). They are sending more staff to prestigious business schools, such as Harvard, to educate them as managers of multi-national corporations.

For European trading companies the most practical way to succeed in the Japanese market is to utilise the financial and marketing capability of their Japanese counterparts which have established marketing and distribution channels for nearly every commodity and product throughout Japan. In fact, there are many such cases of cooperation. For example, Scotch whisky, French wine and Spanish sherry are exported by European trading companies, imported into Japan and distributed by Japanese trading companies.

In the case of products, patents or know-how and the finance related to the new industries mentioned above, European trading companies will find more business opportunities through cooperation with Japanese trading companies which are the core of the organisational network of these new industries in Japan.

Often, European manufacturers or trading companies who are newcomers to the Japanese market hesitate to utilise Japanese trading companies in order to save commission or other expenses. This is a false economy because the special demands made by the Japanese market, such as the complicated distribution system, the different attitude of consumers and the problems arising from the language will be all the greater if a firm attempts to go it alone.

Some guidelines for using trading companies

- First, identify your own needs. Do you need a trading company to buy as a principal or as an agent or distributor in Japan?

- If the trading company is acting as an agent, what functions do you expect it to undertake? Research the following then you can decide on the functions that the trading company can carry out as your agent:- the pattern of distribution channel for your product; the type and location of customer; whether it is necessary to carry stocks in Japan; whether after-sales service will be required and what form it should take.

- You will have to decide whether it would be helpful for your product to be marketed alongside similar products already undertaken by the trading company.

- You will determine whether you will wish to exercise any control over marketing aspects such as distribution policy and advertising.

- *The trading company's group.* All the larger trading companies are members of a group which includes a bank, manufacturers, etc. What advantages are the group companies likely to bring to your product?

- *Financial power.* The financial capability of the trading company is very important when your final customers are medium or smaller factories or wholesalers. The trading companies often have to finance deals with these customers.

- *The trading company's philosophy.* Does your product generally fit in with the company's own explicit development pattern? For instance, some Japanese trading companies have stated that they are interested in promoting the leisure industry; others are connected with promoting supermarkets, others are interested in computers, and so on.

- *The trading company is also turnover-oriented.* They are accustomed to handling standardised mass-sale products and commodities. They are not, in general, interested in handling non-standardised small-turnover consumer products and very specialised capital goods. You must utilise smaller specialised trading companies to market those products in Japan.

- Does the company have qualified staff and expertise to handle your product?

- Seek the advice of a trade organisation such as the local Japanese Chamber of Commerce, the local Japan Trade Centre of JETRO (Japan External Trade Organisation), the commercial section of the Japanese Embassy, or your own local embassy in Tokyo. Alternatively, approach the local office of one of the trading companies and ask them frankly about their ability to market your goods.

20
Joint Ventures

ROBERT J. BALLON

An international joint venture in Japan can rightfully be considered the epitome of doing business with the Japanese. In business there is more than the contract; there is the relationship, be it for distributorship, licensing or plain purchasing and selling. There is also more than corporate control; there is cooperation to be expected from managerial counterparts and from the work-force. And for financial reporting there is an unavoidable cultural aspect to be evaluated.

These business practices impinge on any foreign operation in Japan. The advantage of the joint venture, however, is that it often permits more controlled experimentation, a sort of programmed learning. For example, personnel problems flown in the face of the foreign executive heading a 100 per cent subsidiary, are largely handled by the Japanese partner in the joint venture. But there is a difference: in other types of operations, the lessons of doing business in Japan are *learned*, often the hard way; in the joint venture, these lessons are *taught*.

Negotiating the joint venture

In Japan, marriage has been and still is to a considerable extent less the union of two individuals than that of two families. This tradition is kept alive by the role of the 'go-between' who is officially in charge of the proceedings and who turns what is, or may become, a 'love marriage' into the socially acceptable 'arranged marriage'. Such marriage counsellors in Japan, therefore, are more concerned about the compatibility of the families than about that of the partners.

The same must be said about a joint venture 'business marriage' between a foreign company and a Japanese company. The language barrier and cultural variants call for adjustments between the partners. But 'business bliss' requires more. To put it in a nutshell, the foreign partner is particularly concerned about law and the way in which the new household (company) will be run. Quite naturally, therefore, he concentrates on the 'marriage contract' and all its legal intricacies. On the other hand, the Japanese partner is concerned about the relationship and its results. To the Japanese a joint venture is a *kogaisha* — a 'child company' — which also implies that 'in-laws' should take joy and pride in the offspring.

Negotiations in business, as elsewhere, aim to reach a meeting of minds. Obviously, the prospective partners will discuss their objectives and reach some mutually satisfactory arrangement. The first stage of any joint venture in Japan is negotiating; if this is not done properly, painful situations may arise at the start of operations which can vitiate the relationship for years to come.

Whichever of the two (or more) prospective partners first took the initiative in founding the joint venture, it should be remembered that this factor will always play a part in the future course of the relationship between the parties. Much negotiating skill will be spent in establishing the equity participation: the fact that there may be no choice but the 50/50 solution is no panacea. Finally, the importance of recording these negotiations cannot be stressed too often.

Proposing the joint venture

Exactly who originates the proposal for a joint venture company is not always clear. Two typical case histories are, first, when the idea grows out of a long-standing trade relationship. Such a natural development, as will be seen later, presents psychological advantages; however, it can easily result in a lack of effort by the partners in properly planning their new form of cooperation. This is often the case where the joint venture is entered into with a former agent.

The second example, which actually occurs more frequently than one would expect, results from a personal relationship, such as the casual meeting of two company presidents who, while 'talking shop', discover some common business interests. Since, in Japan, the company president (*shacho*) has an altogether different function and must possess different human traits from those of his counterpart in the West, a meeting of minds even at this level of the management hierarchy may have very little bearing on further developments.

Between these two extremes there is the full range of business motives, including ulterior motives. In some cases, the initiative is largely left to the Japanese side, with the result that the foreign parent company adopts a rather passive stand in the actual management of the joint venture by not having resident foreign managers and staff and by exercising only limited control over such matters as quality and market distribution.

In recent years, the foreign side, for reasons of its own, has tended to take the initiative in promoting joint ventures, either because it wants to establish a permanent position in the Japanese market, or because this has been a condition for acceding to the desires of the

Japanese side. For both partners the joint venture is usually a second best. Surveys conducted by the Ministry of International Trade and Industry (MITI) have revealed that in the majority of cases neither side initially intended to establish a joint venture. Thus, this form of cooperation is often essentially a compromise, frequently glossing over deep-seated incompatibilities.

The 'right' partner

Much concern is always given to finding the 'right' partner. The trouble is that there is usually no clear-cut criterion for determining who is right, except by asking if there is a good chance of reaching a *fundamental* agreement. It is not enough to agree (or give in) on all the detailed provisions of the agreement. As full and complete an understanding as possible should be reached on the motives of each prospective partner before the contract is signed (or for that matter before the Letter of Intent is presented).

Another problem concerns the ulterior motives of each partner, which are of course left unspoken. Such mental reservations damage the joint venture from the start. Trust is not promoted by distrust, neither can trust be engendered by legal documents. To be frank, the very first question which has to be answered is: 'Can a Japanese company be trusted?' The answer is that it can be trusted no more nor less than any company at home. The records clearly indicate that the Japanese is a trustworthy partner; gossip to the contrary should be treated as if you were dealing with a local domestic company. The problem, when it arises goes back to the lack of fundamental understanding; this cannot be assumed to exist merely by virtue of the written agreements between the parties.

What are the ingredients of such an understanding? For the Japanese, it would first involve a long-term proposition and not be the result of some immediate agreement. Major policy changes at the foreign home office, usually decided without prior consultation with the Japanese parent, are rightly or wrongly quickly interpreted in Japan as a breach of good faith.

Secondly, this understanding should be reached at corporate level. Too often it appears to have been obtained only with the foreign negotiator and his immediate superiors at the time. If it is decided to station a resident foreign representative in Japan, this executive should be thoroughly familiar with all the preceding events.

Thirdly, an understanding of the similarities between parents is not enough; it must cover also the dissimilarities, as dissimilarities and not just as points wrapped up in a general compromise. So it is vital that each parent reviews its expectations with the other before any contract is signed.

Joint expectations

It is often supposed that, in view of the capital distribution of the joint venture, conflicting expectations will somehow be resolved in practice. Let us remember that each partner would have preferred to establish its own wholly-owned subsidiary. The Japanese would prefer to acquire the desired technology without foreign equity and management participation, and the foreign side would prefer, other things being equal, to run Japanese operations single-handedly. On both sides these basic preferences may be short-sighted. Hopefully, the negotiations will show both partners that, especially in the long run, cooperation is not a second best but perhaps a first best.

At the risk of some over-simplification, it could be said that the Japanese side is primarily interested in gaining some *technical* advantage, whereas the foreign side is looking for a *market* advantage. To the foreign parent, the new venture appears to be more of an asset, and to the Japanese parent more of a liability. The extraordinary appetite of Japan for importing technology (product- and process-orientated as well as managerial techniques) is no less well known than her extraordinary capacity for assimilating and developing it further.

So with some reluctance the Japanese parent in the joint venture must now consider contributing capital (mostly borrowed), personnel and local expertise to the joint venture. The foreign parent, meanwhile, has learned about the *affluent* Japanese. Rather than contribute to enlarging the Japanese pie in return for a windfall profit in the form of royalties, it now expresses the intention of getting its piece of the pie. It will provide capital and technology in return for a certain degree of control, usually in the form of management participation.

The perfect compromise?

But rather than evaluate mutual expectations, negotiators tend to concentrate their energies on a number-game whose basic rule is 'equity participation = management participation.' Since the initial desire of both partners was 100 per cent ownership, the meeting point is of course the 50/50 joint venture, apparently the perfect compromise. It may even go as far as splitting the joint venture with the same partner into two companies: a 49/51 (foreign/Japanese) manufacturing company and a 51/49 marketing one. The fact is that a true 50/50 position is a deadlock; to resolve it some special mechanism is necessary — such as a third party. In practical terms, therefore, the meaning of the 50/50 relationship is that a veto power is granted to both sides. If veto power is really what the parties want, with proper drafting of the Articles of Incorporation, this can be accomplished under Japan's Commercial Code with *one-third* of the shares. Why then waste so much energy trying to establish it with one-half of the shares?

Another eminently practical aspect of the 50/50 myth is that, if the board of directors consists of six members, three will represent the foreign side and three the Japanese side. But in fact only one of the foreign directors will be a resident. Again, therefore, in practical terms — of daily management, the board is not a 3/3, but a 1/3 proposition. If such a board convenes, it is often either pure sham as the meeting is held nominally and minutes are made up only to satisfy the law, or it is held as a meeting of investors, not of the top management of the joint venture. Such being the case, the management of that joint venture will not be 50/50, but 100 per cent or almost Japanese. This is adequate reason to question the concept of management participation itself unless the foreign side is satisfied with the kind of remote control provided by financial reporting, a problem which will be considered later.

But let us return to the negotiations. In a sense, they are a sort of *dry-run* of the coming relationship. The foreign parent company will probably be represented by some trouble-shooter planning to clinch the deal in a matter of a week or two. The Japanese, for their part, play the game on their home ground and are ready to give the matter all the time it requires.

The foreign negotiator on the spot

The foreign negotiator soon senses that business in Japan is not a study in black and white, but consists of all shades of grey. After a few days of so-called negotiations, he begins to feel insecure, and as a result clings to the written word of the draft contract he brought with him. Rather than negotiate a business relationship, he tends to negotiate a business contract. This is not the best approach in Japan.

The sophisticated Japanese partner will distinguish two basic approaches to the business contract—American and European. The American approach produces a first-class legal instrument laboriously prepared by the corporate legal counsel who, many thousands of miles away, has tried to provide for as many loopholes as he can imagine. The European approach is to consider the contract as a scheme of grand strategy, carefully figured out in the mind of the negotiator, aimed at getting the best possible deal. Neither of these approaches is Japanese.

The foreign representative usually rediscovers what he was told all along, namely that the Japanese are tough bargainers; the puzzling thing that he probably was not told is that they negotiate from a *low posture.* This attitude is not the result of an inferiority complex; it is a characteristic of Japan's *vertical society*. A man is always inferior or superior to somebody else. This may upset western bargaining tactics that usually are assumed to take place on an equal footing. In Japan, negotiations take place as if the two partners were standing on a stairway at two different levels.

As negotiations drag on day after day, the foreign representative is confronted by a large group of Japanese who keep exchanging remarks (in Japanese), take copious notes (in Japanese), shuttle in and out carrying voluminous memos (in Japanese), endlessly repeat questions already answered, etc. etc. The experience is unnerving.

Then one day, quite unexpectedly, the Japanese declare that they are in agreement. Even complex and delicate matters are now settled so easily. And the Japanese are surprised that everything must be submitted once more to the eagle-eye of the home office overseas. Their feeling about this second review is similar to the western negotiator's own feeling about the Japanese second view — i.e. the one taken by the Japanese government under its validation system. More rounds of negotiations are then called for, both sides implicitly agreeing that they are too far engaged to pull out.

Japanese negotiators in the field

The Japanese will always be deeply concerned with what the new venture will mean to their existing organisation and to the close network of relations that characterises the Japanese business world, with the other members of their industrial group, their financial backers and their counterparts in the industry. All these 'relatives' of the Japanese partner, invisibly present at the negotiations, are taken very seriously indeed. This can be seen in the delays to answers, long silences and interjections from subordinates.

Throughout the discussion there will be a constant desire to probe further into the intentions of the foreign side, not out of distrust, but out of anxiety to adapt themselves more fully, and to be ready to react immediately to the coming 'human' problems that are expected from their direct association with foreign interests. The Japanese side, in fact, is not negotiating a business contract so much as a business relationship, and a lasting one.

Record of the negotiations

The foreign partner tends to believe that the ultimate aim of the negotiations is the proposed new company and consequently focuses his attention on the necessary legal documentation. However, at this stage of the game he should also use his 'legal' mind to keep a well-documented record of the negotiations. Often, years later, this record emerges as a vital contribution in the survival and development of the new company. From the viewpoint of the business 'relationship' — rather than the business 'contract' — these minutes have the great advantage of being more personal and more direct. Maybe, after several years of operations, when the negotiators are no longer available, certain policies will regain their substance through these minutes.

A major hindrance to a long-term smoooth relationship will be the quick turnover of foreign executives in the joint venture and at the home office. The record of the negotiations will be of great help to any newcomer not only for his own understanding of what is going on, but also to perpetuate mutual understanding.

★   ★   ★

ESTABLISHING THE JOINT VENTURE

In Japan negotiations are lengthy because they aim at establishing a new business relationship or giving an old relationship new form. Too often the foreign side would prefer to speed up, if not altogether short-cut, the negotiations.

From what has already been said it should be clear that where the foreign side relies on legal documentation, the Japanese side would rather rely on an effective human rapport. Although these two approaches to business are not mutually exclusive, nonetheless there is a clear difference in emphasis. This point can be illustrated by a simple comparison of the numbers of practising lawyers in Japan and in other countries. In round figures, per 100,000 people, the United States has 150 lawyers, Switzerland 42, West Germany 33, and Japan 8. Different social dynamics are at work, not in the sense that Japan has almost no disputes compared to other industrial countries, but that litigation is handled in a different way. The fact is that the Japanese rely less on the legal profession and the courts; even when the courts are called upon, they are valued more for their help in reaching a compromise than in issuing a judgement.

The contract

Can one expect the Japanese to abide by the contract? The answer is not less and not more than westerners. However, the true binding force that the West expects from the contractual relationship, Japan expects to come from the human relationship, the gentlemen's agreement. Some disturbing stories of Japanese default can easily be gathered from any group of foreign businessmen; similar stories about western businesses are available from Japanese businessmen.

Let us turn to the contract. On the Japanese side, legal counsel will mostly be informal, provided from within the organisation of industrial group. Basically, it amounts to a personal guarantee of the partner's intentions. Usually, it is the foreign side that brings in formal legal counsel.

Occasionally the foreign investor expects that all aspects of a joint venture can be set out in a single document. This approach is not consistent with Japanese law and practice; in fact, the foreign party can better secure his position by working out a variety of documents as described below.

The Articles of Incorporation, which must be reviewed by government agencies and registered in the corporate registry, form the basic corporate document for establishing the new company. Its general requirements are set out in the Commercial Code, which can also be relied upon to enforce its terms. Of special importance is the wording of the purposes of the new company, since later changes will require amendment of the Articles of Incorporation and government approval.

Technological agreements, (technical assistance, patent and know-how, licences, distribution contracts, etc.), because of their technical nature, should be drafted separately. These are contractual documents and are enforceable as such: they must be specifically approved by government authorities which may require specific revisions of certain provisions.

Joint venture partners often make a separate agreement dealing specifically with *major policy decisions*. In short, such an agreement specifies certain decisions concerning the operation of the joint venture company that cannot be made without the express approval of both partners. This, too, should be shown to government authorities and will be an enforceable contractual obligation.

It must be emphasised that all these documents should be tightly interrelated and should be presented to the various government authorities as a complete package. Piecemeal submission is dangerous since one matter might be approved and another rejected.

Government scrutiny

It is sometimes said that the Japanese partner is able to agree to anything knowing full well that the government, at the time of validation, will secure his interests. Such cases exist. But basically the reverse is true. The Japanese side, in fact, has to sell the whole idea to the government, acting not only as promoter of the national interest but also as arbiter for the industry concerned.

The application for validation is another of the necessary legal documents for the establishment of the joint venture. The application must include an explanatory statement describing the proposed business of the joint venture and showing how it will benefit the Japanese economy (without having an exceptionally detrimental effect on Japanese industry). The drafting of the application itself has special importance in that the activities of the joint venture will be limited to those matters stated in the application. If the company later wants to engage in new activities, these will require approval through yet another application.

Before the application is submitted, an informal sounding of the government's position is regarded as important, and is very necessary. The prospective Japanese partner and/or the representative of the foreign side will have verbally informed the appropriate ministry departments about the specifics of the plan, so as to try and find out what the official reaction will be. What, to the foreigner, appears as sheer bureaucratic arbitrariness, is, for the Japanese, business participation between government and industry. It is good to remember in this instance that Japan's leading bureaucrats are top-flight intellectuals: the best graduates of the best universities still aspire to this career. At this stage, however, the informal approach will not necessarily result in positive support but it may indicate what hope for approval exists without substantial revision. Formal approval procedure consists in presenting the complete set of legal documents, together with the validation application and the explanatory statement, to the Bank of Japan (BOJ). From here the documents will be channelled for study to the various agencies; they end up at the Foreign Investment Council for formal approval. This approval is then transmitted to the BOJ which will inform the applicant of the result.

Even when official approval is received, there remains one more government hurdle. All international contracts, including the joint venture contract, must be filed within thirty days of their execution with the Fair Trade Commission (FTC) which judges their appropriateness from the angle of anti-monopoly legislation. This legal requirement became standard procedure over ten years ago. The FTC does not issue an 'approval' but simply instructs the parties to delete or amend offending provisions. If the FTC's guidance is not heeded, it may turn to the courts for legal sanctions.

The Board of Directors

The joint venture agreement and the Articles of Incorporation determine the proportionate representation of both parents on the new Board as well as the extent of their ultimate authority over the joint venture. This is the problem of corporate control.

On this Board, a minimum of three directors is required by the Commercial Code (Art. 255); usually the number is six to seven. In Japanese corporations the highest position, Chairman of the Board (*kaicho*), is generally occupied by a former president who has little *de facto* responsibility. In a joint venture, the title of chairman may well go to the president of the foreign parent company who will be listed as a non-active director. It is advisable, at some later date, to replace him by a Japanese national, who then may be extremely helpful in government relations as well as for more delicate personnel problems.

The president (*shacho*) of the joint venture company should be Japanese. (Quite a number of joint ventures have a non-Japanese as president, thereby compounding, at the very top of the company, all the problems encountered by the foreign representative.) At the start of operations, this position is often filled by the president of the Japanese parent company; later he should be replaced by a senior executive of the Japanese side. The president, as chief executive officer, will be one of the two or three 'representative directors' appointed by the Board from among its members to represent legally the corporation (Commmercial Code, Art. 261).

The other Japanese directors, in all probability, will be full-time operating executives in the joint venture, with the possible addition of the top financial officer of the Japanese parent. It is to be expected, therefore, that Board meetings will readily turn into discussions at the level of the operative rather than the corporate policy.

The International executive

On the foreign side, it is fairly common to have several non-active directors, usually executives at the home office, besides the foreign resident representative, who should also be one of the 'representative directors'. His function is most challenging in that he is expected to be the only true international executive in the entire group.

All that has been written about the international executive applies to Japan, and even more so to a joint venture in Japan. More specifically, the following appear to be the main problem areas.

● It has often been noted, in Japan as elsewhere, that the *resident foreign representative* is called upon to assume a much wider scope of authority and responsibility than would be normal at the home office. This scope is probably even wider in Japan, as very often the home office will be in no position — notwithstanding claims to the contrary — to evaluate properly what is going on in the joint venture, be this the result of language and cultural barriers or of the different tempo of economic growth.

● Since *continuity of personnel* is important in Japan, it should be matched on the foreign side as far as possible. Ideally, therefore, the foreign resident should also have been a negotiator of the joint venture; his successor's first task should be to familiarise himself with past events.

● Given the *traditional respect for age* still prevailing in Japan, if there is a choice, preference should be given to a senior foreign representative — perhaps a man in his forties. Should he be able to speak Japanese? Theoretically, yes, but this does not mean he should be fluent. And in practice open-mindedness will be his most valuable asset.

★   ★   ★

MANAGING THE JOINT VENTURE

A new legal entity has now been established, the joint venture company. From the mere legal viewpoint, the joint venture could have been incorporated either as a *kabushiki kaisha* (joint stock company) or a *yugen kaisha* (limited liability company). But the Japanese side insisted on the first form, largely because it has greater prestige in Japanese society, and is considered to represent a more explicit commitment than the *yugen kaisha*. What is this commitment? The answer calls for an explanation of the management of a corporation in Japan.

Kabushiki kaisha, or KK as it is usually abbreviated, is a literal translation of the western legal term 'joint stock company'. But the term stands for quite different economic and sociological values. The clearest evidence is given by the debt-equity ratio of Japanese corporations which averages 80/20. Such heavy reliance on debt would in the West spell bankruptcy; in Japan it means growth. It does not necessarily mean bank control, but very simply *interdependence* of all business institutions, including in the last resort the Bank of Japan. In other words, whereas equity capital in the West is the largest portion of a company's capitalisation and stands for the *individuality* of the corporation — thus individualising the business risk — in Japan, debt capital stands for *interdependence*, thus spreading the risk and stimulating growth on a collective basis.

Interdependence

Outside this network of mutual obligations there is little hope for success, even in the case of an international joint venture company. Since this is the way the Japanese partner and the Japanese managers look at business, even in the day-to-day management of the joint venture, the foreign side should be prepared for interdependence rather than independence of operations.

The first manifestation of this interdependence is that the Japanese parent will consider the joint venture as a *child company*, added to the cluster of companies that form a typical Japanese industrial group. The tendency is then to manage and evaluate the new company in terms of its contribution to the group. Though this is rarely to the liking of the foreign parent, it may well be the secret of long-term success, but at the cost of some short-term sacrifices.

It will take time for the joint venture to reach 'adulthood'; in fact, maturity cannot be reached as long as the practice of personnel transfers from the Japanese parent company persists. But even when adulthood is reached, it will be less as an independent company than as a responsible member of the industrial family. Its individuality is not denied, but its strength is derived from participation in the group. Unless this fact is recognised by the foreign side, the company will remain unfulfilled, to the frustration of its own work-force which will feel outside the mainstream of business activities. The absence of participation in trade associations and management organisations, the lack of managers who enjoy social standing in the industry, failure to cultivate contacts with the banks and the government agencies, etc. — all these have a demoralising impact.

Personnel problems

One of the striking features of business participation is the common practice of transferring personnel from the parent to the child company. Given the lack of labour mobility among corporations, this is a necessary outlet for taking care of surplus and aged employees. One of the biggest hurdles facing new businesses in Japan is the recruitment of trained and senior staff, since, traditionally, Japanese companies take their new employees direct from school and university and train them from the start. The initial staffing, therefore, has to be by transfer from the Japanese parent company or from within its group.

Qualified staff as a rule will be transferred on a temporary basis, say, for two to three years; the foreign partner, who discussed staffing policies during negotiations, should insist that at least some of the managers of the joint venture be transferred permanently from the beginning.

Since the joint venture approach is usually a second-best alternative for the Japanese partner, the danger exists that he will not necessarily volunteer his best people. There will also be some managers, close to retirement age (in their early fifties), not really needed by the parent company. and available for a higher position in the joint venture. This is a standard Japanese form of recognition for many years of loyal service.

Manufacturing

Joint ventures in Japan are essentially manufacturing and/or marketing operations. But remember that the lure of the numbers game may have brought about a 49/51 equity distribution in manufacturing and a 51/49 in marketing — in other words, two contracts governing one relationship.

A manufacturing joint venture company often appears much easier to handle than a marketing tie-up. The main reason, perhaps, is that manufacturing requires fewer value judgements than marketing. As noted earlier, the Japanese are eager to acquire foreign technology and to apply to it their own ingenuity, especially for further development. It has often been said that Japanese Research and Development puts the stress on D rather than on R. Such statements are based on two dangerous clichés.

(a) 'The success of Japanese development in R & D is due to their talent as copiers.' In recent years, it has become increasingly obvious that 'development' implies much more than 'copy'.

(b) 'The weakness of Japanese research in R & D is due to an innate creative poverty.' Again, in view of recent happenings, it would be safer to assume that the Japanese simply felt that it was cheaper to 'buy' new technology than to 'create' it. It seems that national pride noes not deter the Japanese from assimilating and thriving on western technology.

The foreign partner probably will have little difficulty in negotiating favourable terms for transferring his technology. He would also be advised to negotiate cross-licensing rather than straight licensing, knowing the Japanese propensity to improve on his technology. Without such cross-licensing it soon becomes difficult to determine the original foreign contribution.

Once the contractual arrangements for the transfer of technology have been established, the entire manufacturing process is left largely to the Japanese side. Though the Japanese are very eager to learn *all* about the new techniques, the transfer may well be hindered if the foreign technician also acts as manager. Proprietary protection should not be sought through so-called 'technical assistance'.

Personnel problems in a manufacturing joint venture are easily complicated by the presence of a blue collar work-force. Initially the work-force may have been provided *en bloc* by the Japanese partner. Such transfers, however, should be permanent. Again, this point should have been carefully discussed at the time of negotiations so that agreed principles can be invoked. The idea here is that when faced with some labour problem, the joint venture company should not have to call directly on the Japanese parent company, but be in a position to invoke its own principles of personnel administration as established at the time of the negotiations.

Manufacturing will almost always bring the foreign partner into direct contact with the widespread practice of subcontracting. In Japan, this is a close business relationship. Much of the business participation dynamics, described earlier, enters into play. In the eyes of the foreign manager, too much will be decided on a subjective basis, apparently to the detriment of economic rationalisation. This attitude proves that the foreign manager does not yet appreciate the multiple ramifications of inter-dependence.

Marketing

Most joint venture companies in Japan are in the marketing field. Marketing, as experience shows, is a business activity where the relationship with the Japanese partner comes under greater strain.

The problem comes to a head over the question of adopting more rational marketing procedures, by shorter distribution channels or even direct sales. Agreed, notwithstanding the strong conservatism in the Japanese distribution system, it is always possible to shortcircuit it. Some successful cases, always the same, are quoted. But this is at a cost, and a heavy cost — a point not so often made.

Though rapid changes are apparent, most Japanese corporations are still reluctant to streamline distribution for two main reasons, both related to cost.

Wholesalers (primary, secondary, if not also tertiary) are actually a sales-force that the manufacturer or importer would otherwise have to carry on his own payroll permanently; sales representatives, too, are hired for life.

Wholesalers also perform a financial function since financial transactions are settled by promissory notes. If concentrated in the hands of the manufacturer or importer, this financial burden would soon become intolerable.

Furthermore, distributors also are members of the industrial family, and should not be bypassed simply because of economics. Some foreign interests have earned the reputation of being ruthless — paying a price out of all proportion to the gains by summarily eliminating the intermediary.

Japan's distribution system is under pressure from at least two sides: the manufacturer wants more direct access to the consumer, and the consumer questions the costs of too complex a distribution system. The government is also in the picture with substantial outlays aimed at an overall revamping of the system. One particular aspect of this growing government participation is the increasing influence of the Fair Trade Commission (FTC) on domestic business, especially marketing. The primary function of the FTC is to *regulate* business and trade practices; it is reinforced nowadays by the beginning of consumerism. Its guidelines have started to spell out in increasing detail various facets of marketing in Japan. They are necessary homework for the foreign marketer.

Managing a sales-force in Japan has some peculiar aspects, mostly sociological. A straight commission system is out of the question since the salesman is also a 'permanent' employee. Incentive payments vary in importance in relation to regular pay. Nevertheless, such incentives are usually more group-oriented than individual-oriented. To single out a successful salesman may actually embarrass him, and thus reduce his performance, while jeopardising motivation among his colleagues.

Advertising

Finally, is the question of advertising — the cost of which is often of staggering dimensions. Again, it is interesting to watch foreigners compare the cost of advertising and forget that Japan's national dailies have a circulation of several millions and that television peak viewing time involves up to forty million viewers. Costs are calculated accordingly. Savings are then attempted by using advertising campaigns that have been successful in other countries. The risk is rarely worth the saving. Many foreign managers have fallen for the temptation of evaluating Japanese copy on the basis of a translation in their own language. This mistake will be more costly than the fee of a professional copywriter.

CONTROLLING THROUGH FINANCIAL REPORTING

Although legal instruments are meaningful in the case of a major business problem, they are not of much direct help in the control of day-to-day business operations. For this type of control, one turns mainly to financial reporting. What the foreign manager needs here is not only an overview of historical costs, but an insight on today's operations: both are difficult to obtain in Japan. Since the Japanese use the same arabic numerals and since it is easily assumed that accounting terms, once translated, are clear, the foreigner feels that the financial report is his only true and valid 'working' document. This impression is further reinforced by the home office clamouring for monthly financial reports and other financial studies.

Accounting and auditing procedures in Japan are changing fast. The trend is towards complete adoption of the American-inspired Generally Accepted Accounting Principles (GAAP). This cannot be achieved by a mere legislative fiat, notwithstanding the impatient urging of the Ministry of Finance. The small number of professionals in the field remains a major bottleneck; Japan has only about 5000 certified public accountants of whom less than 4000 are active in industry. This number is a small fraction of those of other advanced industrial countries. In Japan accounting has yet to achieve the professional status in enjoys in the West.

Accounting departments in Japanese corporations are staffed from within the organisation; the accountants are people with a great deal of experience but too little formal training. The joint venture company has little hope of recruiting from the open market; it may well have to rely exclusively on transfers from the Japanese partner's staff.

Auditing

With regard to auditing, the 1974 revision of the Commercial Code drew a line at one billion yen capitalisation. For companies below that amount, the function of the statutory auditor (*kansa yaku*), until then largely a sinecure, has been made more responsible. Regrettably, however, the law cannot bestow on him the needed technical capacity. Companies capitalised at that amount or more, similarly to listed companies, have now to submit financial statements audited by independent certified public accountants. In 1974 the principle of consolidated financial statements was acknowledged for progressive implementation (with the hope, meanwhile, that the number of professionals will increase). Currently, such statements have to be submitted to the authorities as supplementary information; but they are not acceptable for tax purposes.

Ironically for the international joint venture, the auditing function may fare better than the accounting one. Steady strides have been made, since their establishment in the late 1960s, by the so-called auditing corporations (*kansa hojin*) which are similar to western auditing partnerships. It will be no problem, therefore, for the joint venture to be audited by a foreign-based or Japan-based auditing organisation. Audited international reports usually give two sets of figures; one set follows the Generally Accepted Accounting Principles, the other the requirements of the Commercial Code and the tax authorities. Often there are major discrepancies between the two. The reconciliation of two such different sets of figures cannot be done only on paper: it is a matter of developing an understanding of the differences, a major headache for the foreign manager.

Financial statements

However there will be some difficulty on the foreign side to appreciate the financial position of the Japanese partner himself, and to get the Japanese staff of the joint venture to appreciate the nature of financial statements in 'foreign' terms.

The term 'financial statements', used without qualification, will generally not be understood by the Japanese businnessman in the sense of statements prepared for investors and externally audited. He has in mind the traditional statements required by the Commercial Code (the 1974 revision is not yet part and parcel of this tradition); these are much less informative and are not really reliable from the investor's viewpoint. After some perfunctory examination by the statutory auditor and approval by the shareholders, these statements start out as a 'first' set of net-income figures. On this basis, tax returns must be prepared. Since some allowances and reserves for tax purposes are not authorised by the Commercial Code, but are specifically required to be entered in the business accounts, these are now incorporated in the first set and in due course will be approved again by shareholders. Thus they constitute a second set of net-income figures, from which taxable income will be derived at various levels, all local taxes being computed mainly on the basis of the national corporate tax.

If the corporation is publicly listed, it must prepare a third set of figures in accordance with the Security Exchange Act. This was enacted under American influence in 1948 and was thus little in accordance with the reporting standards of the Commercial Code — itself of German origin. One of the purposes of the 1974 revision was precisely the reconciliation of these two pieces of legislation along the lines of the post-war one.

The foreign partner is legitimately bewildered. But so is the Japanese partner who is required to prepare yet another set of figures, this time for international purposes. The simplest method is to agree on external auditing by auditors familiar with both systems of reporting.

This shared bewilderment may develop into near panic when the foreign home office demands financial data which will satisfy management information needs, especially in a format determined by international needs. Such procedures will include monthly reports, periodical budgets and other financial information of varying contents and frequency.

Divulging data

But in should be clearly stated that, under normal circumstances, the Japanese side is not really against divulging financial data; the problem is that it cannot do it in the form prescribed without a change of outlook. Foreign reporting requires a high degree of discernment, whereas in Japan independent judgement — in the sense of a given individual taking full responsibility for some action — is not a part of the business tradition and it is not yet supported by what the West would call recognised professional standards. Furthermore, human elements in the business are still paramount in Japan, as manifested in trade practices that work havoc with accounting procedures: monthly billing, collection by the sales staff of monies which are due, lump-sum payments on fixed dates, customers' open accounts inspection receipts, payment in cash rather than by cheque and so on, as well as the universal reliance on promissory notes and the multiplicity of bank accounts.

In the field of international business, the Japanese challenge has grown considerably, and every indication is that the challenge will continue growing. The challenge is not just to enter and survive in Japan's domestic market, but also to be *with* Japan in its world markets and technological advance. Some ten or fifteen years back, in order to survive as international businesses, European companies had little choice but to establish themselves in the U.S. market. Today, the same must be said about Japan. But, whereas the U.S. market and business practices were fundamentally similar, in Japan history and culture (not to speak of a very different pace of growth) constitute a further problem.

It is often assumed, somewhat naively, that the main obstacle to entering and operating in Japan is her protectionist government. Even without the Ministry of International Trade and Industry and the rest of the protectionist apparatus, the challenge would still remain formidable. The same problem could be expressed in a different perspective. The foreign company looking to establish itself in Japan might feel that it could best do so by being allowed to open a wholly-owned subsidiary, in other words to be allowed its own way. Even so, it would still operate in the Japanese environment and be managed largely by Japanese, a situation not so different from that of a joint venture company.

21
Business And The Law

SUMIO TAKEUCHI

In general, the international businessman can be assured that the Japanese legal system relating to business and commercial transactions is very near to its counterpart in any other democratic society. Let us move on straight away, therefore, to examining certain basic facts, both current and historical. The first point to remember is that Japan's legal framework is based on the 'civil law' or 'codified law' system — as opposed to the 'common law' or 'law of precedent' (judge-made). The two alternative systems, of course, are commonly practised throughout the world.

In Japan, therefore, all basic laws are codified. In fact, one can buy in almost any large city bookshop in Japan a compact and handy manual called *Roppo Zensho*, which means literally *The Six Codes Book*. The six codes are classified as follows: 'Constitutional', 'Civil', 'Civil Procedure', 'Commercial', 'Criminal' and 'Criminal Procedure'. These manuals usually contain not only these Six Codes but also many other codes and statutes and, in some instances, administrative regulations and rules known as Cabinet Orders or Ministerial Orders. Most of the Japanese codes and regulations of any significance have now been translated into English. The reason why the Japanese legal system is based on a civil law system is simply historical — imported as it was from European civil law countries.

Like many other social and political institutions in Japan, the Japanese modern legal system began with the Meiji Restoration of 1868. Until that date and for the two hundred years preceding it, Japan had closed its doors to almost all foreign contact and communication. The Restoration government needed to establish quickly a modernised legal system in order to stabilise its own governing structure domestically and to reopen diplomatic relations on equal terms with America and other leading foreign countries. To achieve this goal the government found it most expedient to import a ready-made legal system. As it happened, the first to be selected was the French system. France, at that time, in fact, had just modernised and restructured its system under the Napoleonic Code.

Hardly had the painstaking work of adapting the decentralised Napoleonic Code to Japan's new era been completed, when the Japanese administration decided that a centralised government would be more effective in restructuring the nation. Consequently, they shifted their point of interest from French to Prussian institutions in which government was more centralised and more authoritative. So, in simple terms, we could say that Japan's civil law and civil procedure law were westernised after the German pattern, while the *Commercial Code* was westernised largely on the French pattern. In addition, there are aspects of Anglo-American jurisprudence contained in Japan's pre-war legal systems.

After World War II, the basic structure of the government in Japan was completely changed into a democratic pattern and a new constitutional law was established which reflected the Federal Constitution of the United States. In particular, it incorporated the provisions of fundamental human rights such as the privilege relating to self-incrimination and equal treatment under the law regardless of sex, religion, and so on. In turn, these fundamental modifications formed the constitutional basis for reform of criminal procedure and other related laws.

Japanese versions of transplanted laws

Corporation Law

Japanese corporation law, which is a part of the *Commercial Code*, provides for several forms of business corporation. Primary distinction is made between corporations of limited liability and those of unlimited liability. The most commonly used, and, in practical terms the only one worth discussing, is the *kabushiki kaisha* ('stock company') commonly abbreviated as K.K.

A later statute modelled on German law introduced another form of corporation, *yugen kaisha* (limited company) for a closely-held corporation of limited liability, which is often compared to the German GmbH. When this separate law was enacted, the legislative intent was to provide for small corporations of relatively limited capital as well as for corporations held by a relatively limited number of shareholders who would benefit from the more lenient requirements of the statute in terms of disclosure of corporate information to the public and other formalities. In actual practice, however, this form of corporation has not been used as often as the legislature expected.

Small enterprises, such as the local chemist, as well as corporations closely-held by a small number of shareholders, are mostly formed and run as *kabushiki kaisha*. Although the reasons for this are manifold, the primary reason why the *kabushiki kaisha* form is preferred to the *yugen kaisha* form is that most of the difficulties and strict formal requirements designed for the protection of public shareholders are either circumvented in practice or have become almost moribund. The *yugen kaisha*, therefore, has lost most of its advantages over the *kabushiki kaisha*.

To quote an example: the public disclosure requirements in terms of corporate accounting and business records are now practically non-operative, partly due to the absence of any active interest in such information on the part of public shareholders. Shareholders' meetings in Japanese companies, which could technically have power to override management, are usually conducted almost as rituals.

Shareholders' meetings of large corporations are more of a ceremony attended by a small number of hand clappers engineered by the management than an open forum for serious debate on corporate policy and management.

On the other hand, in small companies with a small number of shareholders, shareholders' meetings are seldom held, nor corporate minutes prepared, and share certificates are very rarely issued to stock holders.

It is clear, therefore, that the legislative intent to offer more convenient forms of limited liability company has been frustrated in the process of actual application within the particular environment that is Japan, where critical debate and group discussion are not commonplace attributes of society.

Corporate Reorganisation Law

Another example of the difference between Japanese law and the original 'mother law' from which it was copied is corporate reorganisation law.

The present corporate reorganisation law was drafted and adopted after the pattern of American corporate reorganisation law. One American lawyer recently observed that, in the actual application of the law, the Japanese court seemed to conduct itself in a manner significantly different from any U.S. court. For example he was personally involved in a corporate reorganisation case where his client U.S. company had claim against an insolvent company as one of the creditors. He visited the court several times and, much to his frustration, the Japanese court showed strong reluctance to disclose the relevant information about the company to its creditors (including his client). He left Japan with the impression that the court seemed to look upon the creditors as some kind of enemy, while they were in fact victims suffering an unfair reduction of claims as creditors. In U.S. courts, corporate reorganisation is basically a legal procedure in which the insolvent company normally invites the cooperation of its creditors and allows them to take the initiative in reorganising the company in the interest of all creditors.

This is another example of the original legislative intent of the law, as first conceived in the mother country, being obscured somewhat in the particular application it is given in Japan.

Trademark Law and Custom Office Exclusion Procedure
In line with the universal philosophy currently prevailing in trademark jurisprudence, Japan has also adopted the doctrine of 'parallel importation' of 'genuine goods'. Under the 'genuine goods' theory, trademarked goods from a source manufacturer-supplier in one country may not be excluded by the same manufacturer or its related parties working abroad in another country, on the strength of trademark registration made in that country, in spite of the territorial distinctions of trademark rights.

The underlying philosophy is that, providing the trademarked goods originate from the same source, public reliance in the trademark as an effective means of identifying the recommendable quality and source of the goods should not be undermined — regardless of where the goods are exported and circulated.

Until this doctrine was accepted, Japanese courts and the customs houses would usually grant restraining orders to owners of trademark rights, so as to prevent the importation of trademarked goods through unauthorised distribution channels into another country. However, when the Japanese Ministry of Finance decided to encourage the import of foreign products as a means of rectifying the trade inbalance in the early 1970s, the Ministry issued new ordinances relating to customs procedure. These abolished the conventional customs procedure enabling trademark owners to exclude unauthorised importation of their products in an effort to minimise and eliminate the import barriers set by the territoriality of trademark rights. In so doing, the Ministry reportedly drew upon the theory of parallel importation.

In the U.S. and other countries, this doctrine was employed as a means of preventing the abuse of trademark rights in attempts to monopolise the market. Interestingly, it was used in a somewhat devious way, to encourage and enhance imports for purely monetary and trade policy purposes. Yet earlier, the Ministry had been assisting, if not encouraging, trademark owners to try to exclude the importation of unauthorised goods through the use of customs regulations, presumably to discourage imports for contrary monetary and trade policy purposes.

Another peculiarity of Japanese customs procedure is also worth noting here. As outlined above, Japanese customs regulations provide that the import of goods infringing patent, trademark and other industrial property rights shall be prohibited. To foreign lawyers, particularly American attorneys with experience in this field, it might look, at first sight, as if the Japanese customs regulations provide the owner of a patent infringed by the imported goods with an opportunity to register his complaint. It might also look as if the importer charged with patent infringement would also be given the chance to have his arguments heard.

Under the U.S. International Trade Commission hearings (the U.S. counterpart to these Japanese regulations), formal hearings would be held and both sides would present their case. Under these hearing procedures, the result of the case would be predictable, since the evidence produced and the arguments presented could be studied by both parties and, moreover, there would be accumulated precedents on which legal prediction could be based. This is not so in Japanese practice.

In Japan, the import of patent-infringing products may be barred only by the chief of the competent customs office, exercising his full discretional judgement; and the owner of the patent infringed is allowed only to offer information to prompt the customs house chief to exercise his power.

If the petition is accepted and importation is banned, the aggrieved importer can resort to administrative complaint procedures, as may any other person dissatisfied with administrative measures. However, the petitioner-patent owner will have no further recourse in the customs office procedure if his petition is not granted. In fact, the officer would not even need to refuse the petition, because in almost all cases where the official in charge is not convinced of the *prima facie* case of the patent owner, he will simply refuse to receive the petition paper filed by the patent owner and the case will simply be non-existent, *ab initio*, or he will suggest or give 'guidance' to the petitioner to withdraw the petition, in which case it would likewise become non-existent, *ab initio*.

This subtle mode of disposal of administrative petitions and demands through 'non-reception' of documents by administrative agencies is akin in approach and spirit to 'administrative guidance', a well-known administrative technique used by government agencies in Japan.

'Administrative guidance' and the concept of the 'Rule of Law'

The restructuring of various laws was undertaken as a corollary of the democratic concept and the principle of the 'rule of law' introduced into Japan as part of the package containing the post-war Constitution. The criminal procedure law was restructured, for instance, after the pattern of Anglo-American criminal procedure, based on the concept of 'due process'.

In other areas of law, including administrative regulations, structural improvements based on the concept of the 'due process' were adopted. Even so, complaints that the 'rule of law' does not exist in Japan are often made by foreign businessmen and lawyers. Its place, they argue, is taken by the 'rule of administration'. Administration agencies in Japan often have too broad a discretionary power, and dealings with government agencies are usually handled on extra-legal or pre-legal grounds, rather than on strictly-written legal rules and regulations. Administrative intent is usually conveyed through subtle 'administrative guidance', rather than through legal channels amenable to legal remedies.

The genesis and *raison d'être* of this administrative guidance is in the origin of Japan's legal system. Since much of the legal framework, including administrative rules, was transplanted from alien soils, without the accompanying details of sub-rules and long experience, the law enforcement and administrative agencies were obliged to fill in such details on an *ad hoc* discretionary basis as they actually administered the law.

Again, the highly centralised and authoritarian government preferred the direct administration of the law by its executive agencies, rather than through the more time-consuming and frustrating processes of the 'rule of law'.

From the point of the overall cost of social control, the administrative guidance approach is no doubt a more economical and effective means of controlling society. Fewer administrative complaints would then be the subject of litigation in the courts, with the resultant saving of general and specific costs to be borne by the government (or the tax-payer) as well as the plaintiffs. However, in the long term, this may be the more costly and expensive way of governing. The danger in the absence of the concept of the 'rule of law', as far as business and industrial interests are concerned, lies in the fact that no sensible prediction can be made of the results of certain business transactions, where these involve administrative agencies. Thus rational business planning and operations become more difficult, and this could eventually lead to instability in industrial and trade developments.

Be that as it may, during the past thirty years, the approach based on the 'rule of administrative guidance' seems to have worked relatively effectively and efficiently — certainly as far as rebuilding the Japanese economy is concerned. It might be some time, therefore, before a system based on the costly 'rule of law' is properly ingrained and fully integrated within Japan's administrative processes.

Notion of individual rights

Over the past thirty years, the awareness of the range of individual rights that are supported by law has become quite firmly established in the Japanese mentality. The traditional notion that the Japanese do not like to go to court seems to be losing ground. There has been an increasing number of major legal disputes involving consumer movements fighting big business enterprises and, quite frequently, even government agencies. In quite a number of cases, such as those concerned with serious drug side-effects and atmospheric pollution, the victims have fought very obstinately, and often successfully, against the enterprises involved.

Yet, the Japanese as a whole still seem to have an innate dislike for litigation, knowing that they must continue to live for some time to come (sometimes for the rest of their lives) in the same community as their opponent in the dispute, whether that community is a neighbourhood or social community, a business community or even a family community. It is easier to understand why litigation is shunned if one reflects on the fact that most agricultural people living in rural areas and most company employees under the life-time employment system are virtually tied either to their birth-place or the same corporate employer.

In summary, the following caveats are worth remembering:

- You can generally asume that Japanese law is not largely different from western law.

- Yet, one should not be deceived by the apparent similarity or even identity of a Japanese law with its 'mother' law in another country, because it may be interpreted, designed or function in a totally different fashion or for a different purpose.

- Do not assume that the law will always be applied in the same way, even in similar contexts, whenever it is invoked.

- On the other hand, you should not feel too discouraged or frustrated to know that everything is not strictly regulated by, or in accordance with, written law. In dealing with government agencies in Japan, it is reassuring to know that matters will be dealt with according to administrative discretion: these agencies have their own reasons for administering and interpreting the law in the way they do.

- Remember that the Japanese have been and will be constantly in the process of discovering the West's notions of individualism — protected as it is by a legal framework of individualistic rights. The old Japanese tradition of 'extra-legal settlement' will no longer work, once there is no tight 'in-group' relationship with neighbours and friends.

SECTION VI
ADJUSTING TO JAPAN

22
Setting Up A Small Office

SIMON GROVE

This chapter is not directed at the large company which decides, from the start, to establish a large organisation in Japan; such companies have the resources and personnel to overcome all the hurdles and find the short-cuts for themselves. What I have in mind is the individual or organisation planning to start off with a small liaison-office or simply just wishing to explore the ground. On either count there will be two overriding concerns: how much will it cost and what are the snags?

Nowadays, permission to open an office in Japan is usually a formality with more or less automatic aproval to be expected, once an application has been made to the Bank of Japan. The exception to this rule is the office established for carrying out financial business in Japan, in which case fairly detailed procedures have to be followed before the Ministry of Finance will grant permission. Your local JETRO office will give all necessary advice on the formalities required.

Staff

Having obtained permission, you must decide on staff before anything else, and your first choice must be the head of your office, your representative in Japan, unless of course you are in business on your own account.

Major considerations affecting your choice will include salary, qualifications, relevant experience and existing contacts. One of these may be overriding, but when this is the case, remember the need for a balanced and harmonious approach to everything in Japan, particularly when choosing the rest of the team. Sometimes Japanese custom makes this easy — when, for example, the nature of the work or qualifications requires a young man to head the organisation. You can always appoint an older and more experienced man to act as adviser, which is particularly important in a society where seniority is still so important. I am taking it for granted that the young man is a foreigner; it is difficult for a Japanese to occupy such a position, however small the organisation, much before the age of 40 unless he enjoys some innate advantage of background or is a born entrepreneur and leader, in either of which cases it should be said that he is unlikely to be working for a foreign company, particularly one with a small office.

Japanese staff generally demonstrate above-average loyalty and diligence, but they do expect above-average pay and conditions in today's world, particularly if they are to abandon the familiar security of a Japanese company and work for a foreign enterprise which could be here today and gone tomorrow. It is sometimes quite difficult to explain to everyone in the home office why a 28-year-old Japanese secretary is on a higher salary than a 40-year-old middle-manager in London — difficult, at least, until the home office man comes to Tokyo and discovers that the high cost of living and the strong yen currency more than justify what are, on paper, inflated salaries.

Another caveat: hire and fire does not work in Japan, and anyone who thinks otherwise will, after a time, find it difficult to recruit any but the poorest staff. So think carefully before taking anyone on, and remember that it is going to cost you money to part company with him at any time. Quite apart from the established contractual practice of paying a termination bonus to an employee leaving other than unilaterally or because of gross misconduct (and even in those cases, sometimes), there is the expense of recruiting afresh, no simple matter. *Taishokkin* (termination-bonus) is normally on the basis of one full month's pay at the final rate for each year worked; sometimes, and particularly in the case of short terms of employment, it can be more. With such Japanese customs as more or less automatic annual percentage-based salary increments, the considerable cost of employing a Japanese will not escape the careful planner. Japanese companies can partly alleviate this by employing a high proportion of women whom they require to retire on marriage or motherhood; in a small organisation where foreign language and work skills have to be promoted, such regular changes of female staff will not be welcome. For in any small foreign office, whatever appearances are preserved, you will rely greatly on Japanese women. Apart from greater loyalty and flexibilty, you probably have a better chance of recruiting high-grade people amongst women than amongst men. On the other hand, it will only rarely be that you can employ them in senior roles without attracting comment, apart from certain businesses such as fashion or cosmetics.

The Japanese office girl or secretary is used to filling certain roles which are different from those of her western counterpart such as book-keeping, superficial office cleaning (desk-tops, telephones, washing cups , etc. and in fact anything which is necessary to be done in a small office).

Women are the major handlers of (and accountants for) money in Japanese society; the office secretary will often be the frankest adviser in the organisation, as women can stand aloof from career-play and suffer less from oriental considerations of 'face'. In any case, devote just as much time selecting the right secretary/Girl Friday as to every other member of the team.

There are many ways of finding staff, some less sound than others. Human relations being of prime importance in Japan, and your organisation being judged by the company it keeps, you *must* secure staff who will have the respect of the Japanese. You can go to one of the recruiting agencies, foreign or Japanese, which have appeared in the big cities, but you will find their charges high and perhaps recurring. You can rely on the old Japanese network of introductions; the advantage of this is that the introducer, if Japanese, is supposed at least to act as a sort of guarantor. But the introducer's motives need examination; in the ramified Japanese world of obligations, are you getting someone not of great calibre, but to whom the guarantor is obliged?

The best system is undoubtedly the normal Japanese one of advertisement and interview. Depending on the age and experience of the person you want, you advertise in the press (almost certainly the English language press), or through the job-placement bureaux of universities. In the latter case, there are certain formalities required; the would-be employer has to specify to which departments or schools he wishes to direct his search and also to forward full documentation on his company. Over 30 per cent of the Japanese work-force are graduates from some sort of 'university' or institute of higher education: such a high percentage *must* mean uneven quality.

On balance, I recommend the press advertisement, as you are aiming at people who are consciously considering foreign employment and who may have suitable previous experience and certainly have confidence in their language ability. But whatever your means of attracting candidates, there is a well-proven Japanese selection method. (In the case of large companies, there is also an academic entry examination, which need not concern us.)

First, there is a standard *curriculum vitae* form available at stationers and called a *rirekisho*. It tends to be exhaustively thorough — every stage of education from kindergarten appears — but it does tell you a lot. Call for copies in both English and Japanese with photo (there is a space for a photo, sometimes ignored).

Next, you should form a small committee to conduct the interview. The committee should first examine all *rirekisho* received; a substantial percentage can then be discarded without interview

(but it is good form to return the C.V. with a polite letter of
rejection). Whatever happens, *always* have at least one Japanese
member, who must be allowed a period of chat in Japanese with each
candidate. However attractive or convincing you may find the
would-be employee, you ignore Japanese advice at your peril.

The interview itself is much the same as anywhere else in the
world, allowing for different social and cultural dimensions.
Discussion of conditions — hours, salary, holidays, etc. — usually
occurs at the end and then only if the situation looks promising.
There should be very good reasons for a large number of
job-changes appearing on any C.V. — this is abnormal in Japan. If
the candidate is unmarried and living with his/her parents, make
absolutely sure at the interview whether the parents excercise any
veto on the proposed employment. Sometimes this is used as a
negotiating ploy, brought up when nearly everything else is settled;
at other times it can be a delaying tactic or a polite way of declining
your offer. But parents in such cases do exercise much influence in
Japan, and are not unknown to be averse to their offspring taking up
employment with a foreign organisation, particularly a small one. In
nearly all cases, this is much less due to prejudice than to concern
about long-term prospects and security.

At the end, you and your committee are left with a handful of
promising candidates, whose relative qualities can be debated. There
may be only one outstanding candidate; even in this case, you should
not rush to offer the job; in fact, you would be well advised to pay for
a check to be run on the person you have chosen.

In Japan, more than anywhere, appearances can be deceptive. It
may be that somebody you know and trust can vouch for the person;
if so, well and good. If not, you owe it to yourself at least to check
with previous employers, and to obtain a copy of your potential
employee's family register from the ward office, a perfectly open
procedure which gives paranoia to many people in the West but not
in Japan. If you have any doubt, pay a detective agency (about 50,000
yen — expensive, but worth it if you have doubts, and perhaps even
if you don't) to investigate and report. As such procedures are
commonly conducted between the families of engaged couples, you
need have no qualms about this.

It is hard to lay down salary scales in so varied a field of activity,
for Japanese salaries are conditioned by such factors as age,
education, gender, type of work, prospects, experience and
qualifications, and the additional factor of working for a foreign
organisation. There are also questions of conditions, such as:
(i) number of days holiday per annum (it is unwise to stray far from
the Japanese norm here; offer two weeks plus the fifteen public
holidays for the first year, adding a day per year's service and
question the motives of anyone who wants more);

(ii) hours per week (the current Japanese legal standard is 45 per week, based on 30 minutes for lunch) to include such aspects as whether or not there are Saturday working, overtime, and payments for same;

(iii) bonuses (payable twice yearly, normally in June or July and in December; once performance-based, they are now more or less contractual and tend to average 4–5 months salary a year — but remember that severance payments are based on basic pay excluding bonus);

(iv) housing and travel allowances (both tax free within limits; housing allowance is theoretically *in lieu* of company accomodation but is usually a symbolic figure of ¥10,000 or so per month — travel allowance is based on the actual cost of a three-month return season ticket from the employee's residence to the office).

It is advisable to start off with a probation period, possibly with break points and salary increases after six months and one year. If break points are exercised by either party, there should be a predetermined rate of severance payment, although the law does not enforce this for trial periods unless specifically agreed. That way, you can both part company amicably if you cannot fit in with each other. Let us take for example pay formulae based on an *annual* rate to include bonus; it is usually possible to negotiate this in a small foreign office — of course you may stick to the separate-sum system if you or the employee wish.

At 1979 rates, an English speaking secretary might cost a foreign firm's office: ¥2.5 million — 3.6 million

A male, recently graduated:	¥2.3 million — 2.5 million
A male, late twenties, early thirties:	¥3 million — 3.6 million
A male, mid-thirties to mid-forties:	¥3.6 million — 4.8 million
A male, 45–55:	¥4 million — 6 million
A male over 55 (the normal retiring age in Japan) perhaps:	¥2.5 million — 3 million

To these, add housing and travel allowance and also the employee's health and security contributions (¥10,000–25,000 depending on salary). In addition, you would be well advised to bank at least a month's salary a year against retirement payment.

Adviser

You are most strongly advised to have a Japanese adviser beside you, who should be a person of maturity, wise to the ways of the world, and capable of expressing your point of view in Japanese terms (not just language) or *vice versa*. Your average secretary or young male employee will not do. (The need for this wise counsellor is noted earlier in this chapter — on your personnel selection committee, for example — and elsewhere in this book.) He (or she — in this case a woman of forty can be much better than a man of thirty) could be your chief employee, or, as is common in Japanese companies, retained on an honorarium and case-by-case basis. The term 'Japanese adviser' can be read to include any person properly qualified, and not just a Japanese national. It excludes most Japanese and includes a small number of foreigners, although even they will in most cases fall back on their own advisers as necessary, as indeed will most Japanese.

Decisions in Japan being consensual, the role of counsel is still well established. Of whatever origins, your adviser on Japan should be of adequate maturity, status, educational background, experience and connection, having an above-average (in the case of a foreigner a well-above-average) understanding of the other side's social, economic and cultural background and, of course, adequate linguistic ability.

Having chosen this paragon, without whom you will do very little business and behind whom you will sometimes hide when things look difficult, make sure he or she is close at hand in all contractual and bargaining sessions even if only to read the faces opposite you. It is your adviser who also remains suitably sceptical about appearances; it is also he/she who checks the antecedents of those you are about to employ or deal with.

Your adviser must make sure that the parties you are dealing with are indeed entitled to rent or sell the property on offer. Property frauds are extraordinarily common in Japan, partly because the Japanese themselves often misjudge appearances and hesitate to question openly anybody with whom they are trying to do business.

Premises

The next problem is the choice of office location. The options open are influenced by financial and geographical considerations. Costs are estimated here on the assumption that the work involved demands a location not more than five miles from the centre of Tokyo. If you rent an office (or any property in Japan) you must expect major up-front expenditure normally in four forms: interest-free deposits, partly returnable, partly not; advance rent (normally one month); agency fees; and 'key money' payable to the landlord on signing and renewing the contract.

The rent itself, and related to it all the above payments, is based on the area which is always measured in *tsubo*, a unit of 3.3 square metres, or more or less six foot by six. Painstaking calculations appear on the contract, together with a birdseye view of the property and a definition of how the area has been measured — e.g. inner or outer wall measurements or taken along the centre of the wall — together with any additions, being all or part of corridors and landing space, common toilets, etc.

This area in *tsubo* is multiplied by a sum in yen to get the monthly rent, on which all other fees and payments are based. This is not, however, the end of your monthly expenditure. There will be shares apportioned for janitor charges and communal services (cleaning, lighting and heating or cooling).

The Japanese landlord still possesses very substantial rights, both on the basis of law and social custom, particularly before he permits occupancy but also thereafter. He can refuse a tenancy at whim, and he can be as prejudiced as he chooses on grounds of age, gender, race, employment or just because he feels like it, and there is no legal or social recourse (beyond the sympathy of friends) available, or even desirable in a society which places more emphasis on avoiding confrontation than on enforcing rights.

There are many landlords in Japan, especially where domestic property is concerned, who prefer not to rent to foreigners, whatever their racial or social background. This is almost invariably because of word getting round about foreign tenants who have 'not understood the Japanese system', frequently by trying to haggle over payments and renewal fees and rates, but also often through the cardinal error of seeking legal enforcement of their rights rather than establishing the harmony of rights and duties which is the Japanese way. As is so often the case in Japan, the tenant must apologise for being in the right before seeking a remedy.

The landlord is far from being an absentee rentier — both embarrassingly and often conveniently, he or his close representative is living on or next to the premises. For in Japan, at least since the establishment of the military peace nearly four centuries ago, there has always been someone required to be responsible for every person or thing. It is, perhaps, by the sum of these responsibilities, more than anything else, that we can define Japanese society.

In return for your discharge of responsibilities, the landlord (in which term is included his representative, the concierge, etc.) assumes multifarious responsibilities to you and for you. Complaints about noise or other nuisance by tenants are referred to the landlord, and are never directly settled between tenants, and it is the landlord who has to sort the matter out. He will maintain records of those necessary details of subordinated responsibility — who, for example, in each office is nominated in charge of fire precautions, the comings and goings of employees (though not on a daily basis) and so forth which the Japanese bureaucracy — police, fire services, etc. — like to know. Your cooperation is needed in this, whatever your feelings about the matter, if you are to reside comfortably in Japan, and a smooth relationship with your landlord must be established from the start and continued to the end.

It will be of some comfort to know that rents tend to fall, in real terms, as buildings age, despite general inflation and rising land costs. This means that, for forward planning, allowance can be made for rent reviews (usually biennnial, with a simultaneous proportionate increase in deposit and a renewal fee of one to two months at the new rental) at below the prevailing inflation rate. Subject to acceptance of this and non-breach of other conditions, there is a very substantial security of tenure.

However, (subject perhaps to the normal *solatium*, depending on the circumstances) tenancies *can* be broken for a variety of stated reasons such as the desire of the landlord to reoccupy for his own enjoyment or that of a close relative. The courts will rule on whether or not the surrounding circumstances make these 'good' reasons, and will more often than not come down on the side of the tenant. The other overriding reason, subject to *solatium*, offer of alternative accommodation etc., is when the old building where you have been enjoying suitably low rents is torn down to make way for another one giving the owner better land utilisation and also higher rents. So, unless you are prepared to move on after two or three years, look into this possibility carefully before renting in an old building for the sake of low rent.

These considerations and details are broadly applicable to housing as well as offices, and in fact the same law, the Housing Lease Law, applies to both. The golden rules at this stage are.

- Do have a contract, or you have poor security of tenure.

- Remember that a lease with no term expressed, *or* a term of under 12 months, *or* yet again one with a term described as 'permanent', 'for ever', 'indefinite', 'in perpetuity' etc., is regarded as 'termless' in Japanese law. This gives you security until either you choose to leave, in which case you need not give notice, or until the landlord serves you at least six months notice.

- In other cases, the maximum term for a leasing contract is twenty years, the minimum is one year. A contract for a stated term of over twenty years is regarded as being a twenty-year contract.

- You will almost certainly be required to produce a Japanese guarantor before entering into a lease contract. You will be familiar with this aspect of life, in that any business or residential visa requires such a guarantor.

Finally, on the subject of office space, how much is it all going to cost? The cost-basis, as noted earlier, is the number of *tsubo* multiplied by a sum of yen, which represents the monthly rent. This sum of yen varies with the location, the age and prestige of the building, the height of the office above ground level, noise, and other factors. Let us call it ¥, and the number of *tsubo* T. Then

T¥ = 1 month's rent

A *typical* example could be as follows:

Six months' advance rent ('*shikikin*', '*hoshokin*', etc.) being an interest-free deposit to cover damage, unpaid rent, etc. Recoverable on vacation, it is in fact a significant proportion of the finance of building costs = 6¥T

Two months' key money to the landlord ('*kenrikin*', '*reikin*', etc.). It may be only one month, it could be three or more. It reflects the seller's market, has no legal basis (but it is virtually impossible to gain possession without it) and is due for payment yet again when the contract is renewed = 2¥T

Fees to agent. Even if you went to the landlord direct, he will bring his agent in on the deal and entrust him with drawing up the contract. Your share of the agent's commission = 1¥T

One month's rent payable in advance on signature of the contract = 1¥T

Total 10¥T

Plus one month advance on common service charges, garage or lock-up space, etc. which, from possession onwards, will be added to ¥T to form the monthly rental payment. Typical rates per *tsubo* per month for office space in Tokyo in 1979 varied between ¥8,500 and ¥20,000. Anything above or below these figures is outside the average.

The office is now yours, and you have your staff; now you must begin housekeeping. The cost of furnishing can vary so widely depending on the nature of the business, the offices, and the demands of your staff and yourself, that only the broadest rule of thumb can apply. An average sum of ¥40,000 per *tsubo*, *excluding* carpets and curtains, is about the bare minimum: beyond that, you can go to really heavy expenditure if you wish.

If you partition the office into subdivisions (essential if you have a telex in regular use, for example) allow about ¥100,000 per *tsubo* of subdivision.

Finally, communications. Internal telephones and telex come within the aegis of NTT (*Dendenkosha*) a quasi-official corporation, whilst externally both come under KDD (*Kokusai Denwa Denshin Kyoku*) a quoted company. To have a telephone installed, you first must buy a bond from NTT, which you immediately discount with a broker. The resulting cost per line installed depends on current interest rates; in early 1979 it was ¥98,000.

To make an international call, you either dial the KDD operator covering the country concerned (see telephone directory) which gives a minimum three minute call with subsequent charges based on a unit of 1 minute, or you apply to the KDD for a direct dialling facility which, once granted, permits very brief costed units of time. In this case, if, instead of the International Prefix Number 001 you dial 002, on completion of the call the KDD computer will call you back (in Japanese) and inform you of the number dialled, the time and the cost.

Most KDD offices have telex machines which the public may use and then pay the transmission charge only; there are no operators available, so this is a do-it-yourself service and only available for international telex. If you decide to install your own telex, you can have a model which will only handle international traffic from KDD, or one which will handle both domestic and international traffic from NTT. Once again, a bond is bought and immediately discounted.

It remains to be said that at the initial stage it *is* possible to operate a small office in a hotel room or business apartment although the viability of this declines as the business grows, only for reasons of image. However, this does defer some of the large up-front payments described, and if you or an expatriate are in Japan it combines the cost of office and personal accommodation. Monthly terms are available at most hotels and there are service apartments. A budgetary indication is ¥200,000 per month. By hiring part-time staff through agencies and using the hotel telephone and the KDD telex, you can begin business with low capital investment and overheads tailored to budget.

But reality is harsher. The average basic cost to international companies of maintaining a small office in Tokyo is currently of the order of £50,000 per year, and the overheads (communications, entertainment, running costs) soon swell this to an average of about double that figure.

23
Smiths Industries:
A European Case-Study

JONATHAN RICE

Smiths Industries is a U.K.-based multi-business organisations with subsidiaries in many parts of the world, employing some 20,000 people and having an annual turnover of some £300 million. It manufacturers, in the U.K., America, Australia, South Africa and elsewhere, a diverse range of products, from automatic pilots to kitchen timers, from marine radar to medicine plastics, from engine instruments for the Boeing 747 to anti-freeze for the family car. Yet the rules for setting up a presence in Japan are basically the same for a reasonably large organisation like Smiths Industries as for a giant international conglomerate or a one-man operation.

The basic and most obvious rule is to study the market before making any commitment. Smiths Industries has had close connections with Japan over many years, beginning in the last century when it sold technical information to the late Mr Hattori to help him establish the company which now produces Seiko watches and clocks. In the past thirty years the company has built up reasonably profitable markets in some sectors, mainly through the use of agents, and that is why some ten years ago the decision was taken to station a permanent liaison representative in Tokyo. His job was not only to make sure that our existing businesses flourished but also to look out for new opportunities that might be open to Smiths Industries.

For most western businessmen the idea of 'business in Japan' means selling goods made elsewhere into the Japanese market which is widely considered to be among the most difficult in the world to break in to. This, of course, provides the Japan-based businessman with a ready-made excuse to give to the people back at head office when immediate and enormous success is not forthcoming, but it overlooks the fact that Japan has for the past two decades or so become famous more for selling than buying.

It is perhaps not surprising, therefore, that among the opportunities which our liaison representative found for us were many components and complete products which could be sourced by Smiths Industries in Japan and sold elsewhere. This is not to say that our sales into Japan did not also grow, steadily and profitably, over the same period, but the presence of our man in Japan certainly enabled many companies within SI to take advantage of the range of Japanese products that were reasonably priced and of high quality.

After seven years of running a liaison office, in conjunction with the activities of our selling agents in Japan, we looked carefully at ways of improving our position. The mathematics was simple. It was easy to prove that SI would save a significant proportion of the costs of running its liaison office by forming a limited company in Japan. Our business in Japan was then at such a level that we could show that by taking over the purchasing activities of all SI companies who were buying from Japan, and by taking over as sales agents for those SI products which were already having some success in Japan, we could live on the resulting commissions, even if sales in either direction did not increase. In other words, it could be shown that to establish a company in Japan could mean for us no further investment of funds in Japan as compared with the cost of a liaison office.

It is opportune now to note that such simple mathematics does not work, although it could have done so if we had been content not to expand our business at all. Thus the second basic rule of setting up a business in Japan is that it cannot be done on the cheap. Investment is inevitable and should be long-term.

The concept of establishing a company in Japan, even a small trading company such as ours, implies that one is proposing to expand existing business with Japan. (Smiths Industries Japan is a trading company with only a staff of a dozen people and is not involved in any manufacturing operations in Japan. Setting up a factory in Japan may well involve a completely different philosophy.) Expansion of business needs money and extra capital tends to alter dramatically the simple mathematics involved in quantifying the benefits to any company of putting its existing business into a new Japanese subsidiary: it is not the existing business that a new company is really established to handle but the growth potential of that business.

Our experience has been that raising capital in Japan to fund the company is not difficult. Smiths Industries Ltd. is of course a public company of some stature with an excellent track record. Nevertheless, it would appear that raising funds in Japan is in general easier than raising funds for a similar venture in most other countries.

One of the first jobs to master was how to say no to the many banks whose representatives visited the newly established SI.J. office to offer money at very competitive rates. The catch is that banks in Japan like to retain an interest in, one might almost say an influence on, the way the business of their clients is run. The relationship between the banks and their clients is far closer than the same relationship in most western capitalist economies. Japanese banks are major shareholders in probably all of the largest Japanese corporations and many of the lesser ones. The debt equity ratio of companies in Japan, encouraged by the banks, is commonly above 10:1 and one soon learns that Japanese industry is founded on one thing only — confidence. This is to a lesser extent true throughout the world, but in Japan a bank will continue to support a company (and of course to influence it) even during a period of poor trading results provided it is confident that the company is basically sound.

A European businessman might believe that it is impossible to say how a bank evaluates the basic soundness of a company, but it is interesting to note one case of a fairly large company which by western standards had been bankrupt for years but which only had the rug pulled from under its feet by the bankers when its debt ratio had reached 98:1 and its chief executive was indicted for a criminal offence.

Our main difficulty has been in adapting our English practices to the Japanese way of life, and once that was achieved, in convincing the keepers of the purse-strings at our headquarters that the way we are operating is not only sane but also normal in Japan. It follows logically that any company contemplating setting up in Japan should put on its payroll as soon as possible a qualified Japanese accountant. This will ensure, with luck, that one problem we faced will not occur. When SI.J. was established we negotiated a loan facility with one of the main Japanese banks. This was easily enough achieved but we felt it was important that a small portion of the loan should be available in the form of an overdraft. After three or four meetings with the bank we were confident the bankers understood what we were asking for, and after another meeting it was readily agreed that a higher rate of interest should be payable on the overdraft as compared with the loan. One more meeting was needed after that for signatures on what appeared to be reams of forms relating to the overdraft facility and then all was set. Or so we naively thought.

One Friday we were due to pay a bill for some millions of yen which would, we knew, leave us overdrawn in the bank by ¥1200 or so. We knew that money to be received on the following Monday would more than cover it, so a cheque was drawn to pay the bill and no more was thought of it. On the Saturday morning our general manager was telephoned at home by a frantic section chief (*kacho*) at the bank saying we were in the red. 'I know,' he replied. 'We have an overdraft facility to cover it.' It appeared, however, after ten more minutes of somewhat heated discussion, that there were another couple of forms we should have signed, not to mention letters explaining what we were about to do that should have been written, before we should have used the overdraft facility. The solution on that occasion was for the general manager to go to the office, to take ¥1200 from the petty cash box and to take it round to the bank to balance the books before the bank closed for the weekend.

The problems of financing a company and then controlling those finances is the most crucial aspect of setting up in Japan. Other factors require common sense but little else. Office accommodation, as noted elsewhere in this section, is not impossible to find although it pays to take into account likely growth over the first few years before deciding on the size. Space is very expensive in Tokyo, but so is moving office too often. The legal side of setting up requires the help of a competent business lawyer. When we established our company in 1976, the legal processes took less than four months from start to finish.

Does a company establishing a subsidiary in Japan need a man from head office to run it or can a Japanese manager be appointed? Our experience is that a man from head office is essential. One of the problems of being in Japan is that of explaining Japan to head office. A man with no experience of his own head office is at an immediate disadvantage and there are times when the head office representative's main function in Japan is that of go-between, making sure that the English and Japanese ends of the business have at least some idea of each other's intentions. This is on occasions a crucial function.

The only other major hurdle which must be mentioned is that of finding staff, a subject dealt with at length elsewhere. I would make the point that finding young staff, clerks, secretaries and switch-board operators, is no real problem. It is the search for competent managers that yields depressingly few who are likely to be any good at all. Beware of the Japanese associate company that says, 'Have one of our top men. Mr Watanabe is a good man.' If Mr Watanabe was a good man, no company, however friendly to the newcomer, would want to give him up. There is no easy answer to the problem, but the only advice is never settle for second best. Once you employ a man who is only second best, he will be with you, in all his stunning mediocrity, for life.

Setting up in Japan is logistically a simple exercise. It is not always the best answer to a company's marketing problems in Japan, but any company which sees Japan as a significant future market or equally important as an international competitor, and which does not at least look at the feasibility of a presence in Japan is, perhaps, taking a very short-sighted attitude.

24
Etiquette And Behaviour

GEORGE FIELDS

There are three key words which sum up the Japanese concern for 'style' — *kunigara* (national characteristic), *kafu* (family style) and *shafu* (company style). It is automatic for a western salesman to set out to know something about his clients' personal lifestyle in order to establish a worthwhile rapport. The Japanese word for style, *fu*, however, applies generally to a collective body, the smallest unit being the family. In Japan, a good salesman has a grasp of his customer's *shafu* (company style) — considered the crystallisation of the organisation's traditional spirit and *esprit de corps*. Of course, a newly established company has no *shafu* in this sense but follows the lead of the president's lifestyle — the head of the company household.

It is obviously very difficult, if not impossible, however, for the Japanese to know your particular western *shafu* — assuming you have one. So they resolve the dilemma by simply substituting *kunigara* — the national stereotype. If your behaviour seems to conform naturally to this stereotype, you are fairly safe. Safer, in fact, than trying too hard to act the Japanese way, although this may not necessarily be such a bad thing, since it could create a certain amount of light amusement and be to everybody's pleasure, not embarrassment. The point is that so long as you conform to the stereotype image, you need not worry too much about offending your Japanese counterpart simply because you have not observed strict Japanese protocol. In my opinion, many foreign businessmen are too sensitive about this issue — in the long run it only serves to cramp their natural style and so puts them at a disadvantage in the context of human relations. The Japanese have fairly fixed stereotype images of major nationalities — the Americans, the British, the Germans, the French, the Russians, etc. — and it might almost be worth your while to find out what these are!

Take your name card seriously

It is not difficult for one Japanese to place his relationship with another Japanese in clear perspective. But it is another matter when it comes to a foreigner, and vice versa. The well-known Japanese custom of exchanging name cards is an essential first step. Having said this, remember that titles differ, not only between Japan and the West, but between western countries themselves; for instance, the chief operating executive is termed president in the United States, but managing director in the United Kingdom. In Japan, like the United States, the president is the chief operating officer but, unlike the United States, where he is sometimes all-powerful, the role of chairman in Japan is virtually an honorary retirement position, allocated to a senior company statesman.

Basically, the Japanese have far more titles and rankings than we do and they find the often ambiguous western titles frustrating. An article in the *Nihon Keizai Shimbun* (Japan's leading financial daily) discussed this question very seriously and mentioned the chagrin of a Japanese businessman who entertained a 'vice-president in charge of the Far East', only to discover subsequently that the man had arbitrarily upgraded himself to this title for his visit to Japan but was 'no more than a *kacho*' (division manager reporting to a division head). To the Japanese businessman concerned, this was one of the most serious breaches of business etiquette.

Thus the article went on to warn that the vice-president title was most likely to be equivalent, at best, to that of *bucho* — 'division head'. Now, there is nothing wrong in the title of *bucho*, and it is a very important one; however, one must remember that in the Japanese decision-making process, which is generally a collective one for major issues, he does report to several levels of directors and finally, of course, to the president. The division head, of course, is an important part of the decision-making process but is seldom the decision-maker. While it may seem a trivial matter, the Japanese translation of your title and position should be checked and double-checked to indicate your true position *vis-à-vis* the Japanese management structure, not yours.

Establishing rapport

Entertainment is a universal tool for establishing a rapport between businessmen; but, unlike the West, business luncheons in Japan are a rarity — they disturb the work flow and, in any event, the Japanese hate drinking during the day if for no other reason than the fact that alcohol often affects their complexion. What is more, evening entertainment almost never takes place in the home.

What is left is evening business entertainment, for which Japan is renowned. Throughout recent years, for example, 1.5 per cent of Japan's impressive GNP was spent on (declared) entertainment, which is more than the amount spent on defence. Entertaining is not limited to clients but is also extended to one's employees and subordinates as a social lubricant. Here, the Japanese have a saying, *hito no kokoro wa yoru wakaru* — 'You get through to a man's soul at night'. *Asobi* (literally, 'play') is supposed to be an integral part of the Japanese management system; in a structured society, this provides the necessary outlet for personal feelings. In turn, the superior obtains vital information for the efficient management of his department but must never be offended by what he hears on these occasions.

As a general rule, inter-company business entertainment basically adheres to this principle although it naturally differs in nuance. As in the West, the purpose of wining and dining in Japan is to establish rapport and friendship but it is also to gain information. The information thus gathered may be perceptive rather than tangible, but this does not make it any the less important. Of course many a contract is signed, many a joint venture deal is concluded in this context, but the final stage is almost certainly preceded by a number of 'information gathering (and exchange)' type of night-time meetings — *hara no saguriai* (literally, 'searching each other's stomachs', i.e., minds — but more on this later).

Status and business entertaining

To be able to entertain at the company's expense is a privilege earned by seniority which, in a Japanese organisation, is derived mainly from many years of service. This privilege may be abused at times but it is never taken lightly. Should it ever happen that a junior company executive is entrusted with the task, he will not only lower the status of the company but, worse still, offend the dignity of the guest. As a rule, etiquette demands that you do not personally invite someone senior in status to yourself; the ideal situation is that you invite someone who is one rank or so below you. If a division head is to be invited, it is preferable that a director, *torishimariyaku*, plays host, since this invariably flatters the guest's ego.

In Japan husband and wife entertaining in the business context is still rare, although, if you are a visiting dignitary, your wife will most definitely be invited. The chances are, however, that the Japanese hosts will turn up without their wives and, apart from the *geisha* or hostesses, your wife will be the only female present. If you are entertaining your Japanese business associates, the best way is to entertain in the style that, paradoxically, requires the least protocol — namely the American style cocktail party. Many Japanese businessmen actually dislike these affairs, since they consider that having to stand up to eat and drink is a barbaric custom. Nevertheless, this western tradition is perfectly acceptable since this is your way. On the other hand, if you are a Japan-based branch

Typical seating arrangements at a Japanese restaurant

(a) 'Zen': the most formal with individual tables

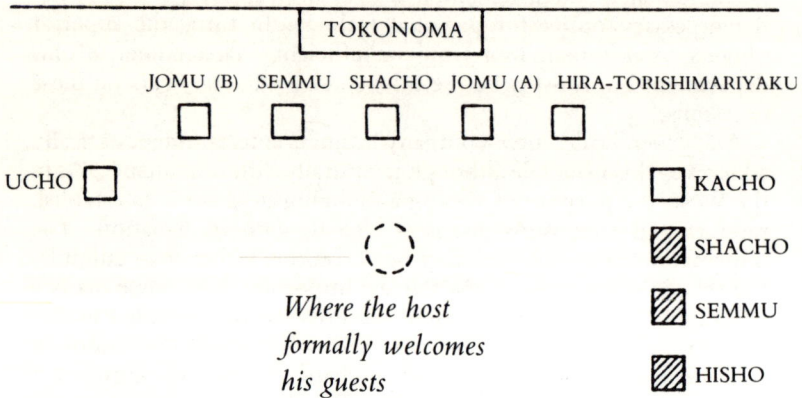

| TOKONOMA |

JOMU (B)　SEMMU　SHACHO　JOMU (A)　HIRA-TORISHIMARIYAKU

UCHO □

Where the host formally welcomes his guests

□ KACHO

▨ SHACHO

▨ SEMMU

▨ HISHO

(b) Less formal with centre table

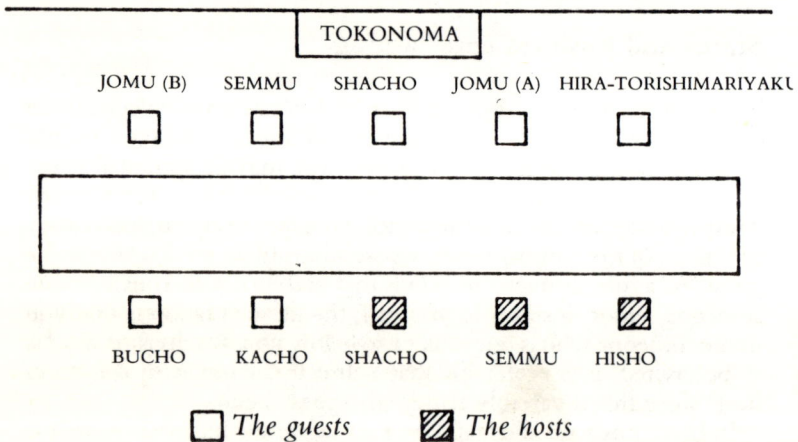

| TOKONOMA |

JOMU (B)　SEMMU　SHACHO　JOMU (A)　HIRA-TORISHIMARIYAKU

BUCHO　KACHO　SHACHO　SEMMU　HISHO

□ *The guests*　▨ *The hosts*

Glossary of terms

Tokonoma	The recess in a Japanese room in which scrolls are hung and flower arrangements placed
Shacho	President
Semmu	Senior managing director
Jomu	Managing director
Hira-torishimariyaku	Director
Hisho	Literally 'secretary', but more akin to executive assistant to the president.

manager or a local representative and you have visiting dignitaries who want to experience a traditional Japanese feast, and you also decide to invite your Japanese business associates along for the occasion, then it would be useful for you to know the protocol of seating arrangements. If you do it properly, your Japanese guests will be highly impressed and will respect the depth of your knowledge concerning their country.

The entertainment ladder

The Japanese word *hashigozake* means 'drinking up the ladder', but it can imply drinking down the ladder — depending on where you fit in to the little hierarchy amongst those drinking with you. Another word relating to this situation is *ichijikai* (the first get-together), then *nijikai* (the second get-together); and remember that no night out can be said to be off the ground and a good rapport established unless everybody adjourns to at least one other bar — the *nijikai* stage. Incidentally, the concept of going 'up the ladder' is not really very appropriate since, as the night progresses, you move on to places feeling increasingly relaxed — roughly corresponding to the general degree of inebriation!

Remember, however, that the first get-together, the *ichijikai*, is a most important occasion and should be approached with the formality befitting the guest's position (ranking). Ignore this, and you are highly unlikely to progress to the *nijikai* stage, and so on.

One word of caution: if, after having been lavishly entertained at the first get-together, you decide to reciprocate at the *nijikai* level, don't use this opportunity to drag your host off to your favourite bar simply because you want to renew acquaintance with your favourite bar hostess. Your local bar contact for these occasions should be impersonal and business-like; a good Japanese hostess understands this and will make sure that your guest is the centre of attention.

Form versus substance

The words *tatemae* (the front face) and *honne* (one's real intention) are the basis of a very important issue in Japanese business etiquette. An entire book in Japanese has been devoted to this subject. Although to the westerner this may seem a cynical form of duplicity, basically there is no difference between the two. The essential point is that 'front face' affords a rule of behaviour which gives continuity from the past and a unified measure to judge the behaviour of others. It is also considered necessary to maintain social harmony and in that sense it acts as an individual's safeguard.

The problem is that these rules of behaviour have actually become increasingly complicated rather than simplified because the social structure itself has become more involved. The individual is not nearly as clearly positioned in society as he once was; thus, the rules of behaviour have become more intricate. Ironically, in this sense, the democratisation of Japan has made the Japanese even more difficult to understand. (Admittedly a heretical view, as most people seem to think that Japan's visible westernisation — instant foods and golf — means that the Japanese are actually becoming western.)

A simplistic conclusion would be to say that the division of *tatemae* and *honne* is an admission of duplicity — the West's raw interpretation of social behaviour. The *tatemae* view is that 'to reveal that making money is bad form' may also strike westerners as a form of business hypocrisy. After all, one may reasonably ask, is there any other reason for being in business than to make profits? The Japanese 'true' answer to this question might be that there is more than one reason for running a business, although he may laugh and agree with you since that may be how you feel. *Tatemae*, therefore, is respect for the other's social style, with the clear recognition that there is very often no *one* right or wrong way. If respect for another's sensitivity is *the* consideration, camouflaging one's true intention is not necessarily duplicity.

In their overseas dealings, the Japanese thought that they were conforming to local standards by not bothering about *tatemae*, which they presumed did not exist. Thus they unashamedly revealed their determination to achieve trading success and felt hurt that they were branded as 'economic animals'! It seems unfair, therefore, that the foreigner has to cope with Japanese *tatemae* in Japan, when the Japanese do not have this problem overseas. It is no consolation, but we have to accept the fact that we are talking about the home ground of the Japanese businessman — and this is how things are done in Japan.

There are Oriental sayings such as 'There are idle moments during busy times' or 'Stillness within movements'. What may seem to be ambiguity to the westerner may not be perceived as such by a Japanese who is irritated (although he may not show it) by the western businessman's noughts-and-crosses approach to strategy and his penchant for analysing everything in logical terms. In the West, being busy means being fully occupied within a time-frame; so the busier you are, the less free time you have. The 'idleness' referred to in the East has nothing to do with such a time-frame and refers to the need to relax the mind — no matter how busy you are. In other words, the mind should be idle from time to time so that you can deal with the situation properly. The point is simply that against this backdrop of basic cultural differences, which are in turn reflected in attitudes, irritations and frictions, are almost bound to occur in East-West business situations. The best business etiquette is not to show your irritation, since difficulties probably derive from cultural differences, and are not a deliberate ruse created by your business counterpart in order to frustrate you and your business objectives.

25
Check Your Own Check-List

MICHAEL ISHERWOOD

For the foreigner the conduct of business in Japan, as has been shown in most of the foregoing chapters, is not necessarily more difficult than elsewhere but, because of the unique nature of Japan, business there is different in many important respects. Here I want to try and summarise the most important elements and provide the reader with a check-list of key requisites.

You need to be well prepared. It is essential to do your homework and to go armed with some knowledge of the Japanese — in a very general sense — as well as the fullest possible information about your own business and the venture you are to propose. The Japanese are always thorough in this respect. They have the benefit of what is probably the finest commercial intelligence system in the world and their education and background experience has prepared them with a good comprehensive knowledge of the western world.

We have a lot of catching-up to do. There are far too few western businessmen who can match the Japanese for knowledge of the 'other side'. This is a statement of fact rather than criticism. The Japanese, after all, have had to learn about us in order to take their place in the industrialised world. Until recently, we have either elected not to know them or we have not needed to know them, but in the course of the next twenty years or so there seems little doubt that the Japanese are going to have an increasingly profound effect on all our lives.

In order either to compete or to cooperate with the Japanese in overseas markets, and certainly in order to enter the Japanese domestic market, it is important for companies to develop a

'Japanese expertise': to cultivate staff, or to set up a department, specialising in Japanese affairs. It is no good sending a different man to Japan each time — it requires men who are long-term Japan specialists, who can inspire confidence and create a sense of personal obligation in the Japanese they meet.

Business relationships amongst the Japanese tend to be highly personalised and, because it is unusual for people to change companies, contacts are maintained through long-term associations. There is a very complex pattern of connections between companies. It is very unusual for a company to stand alone. It is almost always affiliated in some degree to a 'group' with which it works closely and on which it relies for its essential services.

But the word 'group' needs some explanation because it does not have quite the same meaning that we give it in the West. Companies in Japanese groups are not linked together principally by ownership but rather by long-term relationships — a mutual commitment. They are perhaps better described as alliances of companies with varying degrees of relationship and reliance. Holding companies are not permitted under the Japanese commercial code and the element of ownership as a means of control is not as important a consideration as it is in the West.

Japanese people tend to think and act as a group rather than as individuals. This is a basic concept of society quite different from that of the West where emphasis is given to individual self-sufficiency and the development of a strong, competitive personality. In Japan the opposite is required. The Japanese, particularly those in large organisations, are trained to subdue their personalities, to conform to accepted patterns of behaviour and to adopt a conciliatory approach to problems.

The lifetime employment system makes the working group the most important group in life. After school or college the new employee joins such a group, roughly corresponding to a department of a business section and comprising about ten to twenty people. He will be closely connected with them for years after and, because he will always remain with the company, he will personally identify himself with the progress and success of his particular group and the larger organisation of which it is part. In return for being looked after and protected he gives complete loyalty and devotion.

To appreciate the underlying ethos of Japanese business practice, one must look back to the pattern of work and social organisation in the isolated village communities of feudal Japan. Each small community was centred around its place of work and was self-sufficient and inward-looking. People rarely left the village in which they were born and under the Tokugawa administration passes were needed to leave one's district. Each village had a formal hierarchy in which households were ranked according to seniority and status. Seniority depended largely on length of residence within the community. The Meiji reformation simply transferred these principles and practices from the country into factory and office.

Because the pattern already existed in villages, people tended to spend all their time together both at work and play. People you share time with become your friends and loyalty to the group has to be constantly demonstrated. It is a vertical society, several groups making up a department, departments a division, divisions a company. The individual sees himself as part of a group based upon work. Companies do not hire people to fill a specific job vacancy but in anticipation of the need for more people in the future. Skills not provided by general education are achieved through internal company training 'on the job'.

The stability of the working group takes precedence over everything. The rule is to avoid any unnecessary embarrassment or offence to others, particularly within one's own group. Harmony in human relationships is a recurring theme amongst the Japanese. For these reasons the Japanese will avoid direct confrontations or disagreements at all costs and will always seek ways to conciliate. Failure to compromise is seen as weakness and an embarrassing loss of face. This ethic sees expression in Japanese legal practice where courts tend to seek agreement and compromise between parties rather than pronounce judgement.

Since it is almost impossible to express unqualified refusal in Japanese, foreigners are frequently confused by the Japanese inability to say no. In fact, people will often appear to say yes when they mean no. Western businessmen, experienced in the ways of Japan, learn to read the negative response signs — hesitancy in speech and facial expression or an unwillingness to be more specific. But the unwary can be easily misled, sometimes with serious consequences.

Authority is not imposed from above in any remote way so there is little resentment and no social division. Management and labour have the same salary structure based largely on length of service. Pressures tend to come not from above but from colleagues in the work group. Workers have a much greater say in how their work is organised than in the West. The essential characteristic of Japanese management is that it is management from the inside. Managers are developed by the company and are committed to it for the whole of their working life; they are never hired from outside. In effect managers are family heads who receive respect and loyalty from the work group with whom they are closely and intimately involved.

Problems arise when it is assumed that because Japanese companies seem to be organised along similar lines to those in the West (they have, for example, a formal management structure that looks encouragingly familiar) they actually operate western-style. In reality they work through their own unique vertically-structured group system.

The role of senior management is to act as mediator between the various groups (departments, sections) and to coordinate activities in accordance with company policy. They are more involved with personal relationships than with business expertise and it often surprises westerners that they lack the detailed technical knowledge of their western counterparts.

The differences in decision-making can best be summarised by saying that in western organisations most of the major decisions are made at senior executive level and are imposed downwards. In Japanese organisations, the decisions originate at the lower levels where day-to-day business is done and are then coordinated and approved by senior management.

The process of decision-making does take a long time and can prove very frustrating for westerners engaged in business negotiations. The reason is that everyone (and every department) who will be concerned in or affected by the final outcome must be consulted and his opinion taken into account. The Japanese have developed a remarkable ability to modify individual opinions in order to reach a consensus. In practice, when this consensus is achieved, senior management formally approve the decision and everyone will then give the matter their fullest support and the implementation will go ahead very quickly and smoothly. The speed of action after the decision-making period often fully compensates for the slowness beforehand, and often astonishes western observers.

Advice from someone with a good knowledge of the Japanese market is invaluable and it often pays to use the services of a good management consultant or market research organisation. There are a number of very good American and British consulting and research organisations in Japan who can be located through the Embassies or trade organisations. The foreign trading companies should also be considered — they have not the same influence as the big Japanese trading companies but they are easier to communicate with and they have a great deal of experience in the market.

I must stress the importance of introductions when approaching the Japanese. Writing a letter to a company as a form of approach is unlikely to achieve a result — probably even a reply. Letter writing is not a strongly established custom — most Japanese companies are not organised to reply to letters or to unannounced personal approaches. Business is done on a peculiarly personal level and long-term relationships and connections are everything to the Japanese. It is necessary to make your approach through the introduction of someone already well-known to the company you wish to do business with. In the complex network of connections that exists in Japan, you have to remember that the person or organisation introducing you is required to vouch for your reliability.

Essential dos and don'ts

1. Exchange business cards — a formality as important as shaking hands. Go to Japan armed with an ample supply, about two or three hundred. They should give the full address of the company and your position in it. (In Japanese companies titles indicate rank rather than function.)

2. Introduce yourself by your own family name. Avoid the use of first names even if invited to. Japanese always use the family name when addressing one another (except in the close family circle) and first names or nicknames tend to be an embarrassment. If you want to establish a more friendly basis to a relationship that is developing well, add -san to the end of the family name (eg Matsumura-san).

3. Remember that you will always be faced with several people whenever there are discussions; everyone will need to be convinced of your proposals. There will also be other people to convince whom you will never see, so any prepared information you can offer will be much appreciated. It is useful to include information of a general character about your company such as the company brochure, annual report and information about products.

4. Invariably, you will find one person on the Japanese side who will do most of the talking, either because his English is the best or because he knows most about the subject under discussion. However, remember to identify the senior man in the group and occasionally acknowledge his presence.

5. Get into the habit of speaking clearly, slowly and for not more than a minute or two at a time. Emphasise your main points by expressing them in several different ways so that the meaning is clear. Use words and expressions that are simple and avoid slang expressions.

6. The Japanese are apt to remain quiet while they mull over what has been said and what alternatives are open to them when they next speak; they may remain silent while they wait for others to reach conclusions they have come to already. Westerners usually find such pauses acutely embarrassing and feel obliged

to say something to relieve the supposed tension. At best, what is said may appear ridiculous; at worst, you might make a quite unnecessary concession in the belief that this is the reason for the silence. If the silence is caused by the difficulty of solving a problem, the Japanese will happily postpone the meeting to give everyone time for further reflection.

7. It is of course impolite to interrupt someone who is speaking. Foreigners often beome impatient at the slowness and difficulty with which the Japanese speak foreign languages and find it difficult not to butt in.

8. Make the most of your interpreter if you are using one. Brief him beforehand and give him any notes you may have on the proposals you intend to make. Allow him plenty of time to make his own notes during discussions and to clarify points where he thinks the meaning is obscure.

9. The Japanese are reluctant to do business with people they dislike, regardless of the attractiveness of the deal. High profits are not their chief priority; stability, sustained growth and good personal relations come first. It is, therefore, of the utmost importance to establish a harmonious and trustful atmosphere in business negotiations. The Japanese use meetings to sum up people and to gauge the desirability of long-term relations, so your reactions to situations will be keenly observed.

10. Entertainment in Japan plays a major role in any business relationship. When offered it should be accepted and, in due time, reciprocated. Golf should be encouraged whenever possible. It is very popular and is played at all levels of the company hierarchy.

11. Strict rules of etiquette make public interaction easy, predictable and trouble-free in a grossly overcrowded community. Bowing etiquette is most common; the junior person bows lower and even bows on the telephone. In the lift everyone faces front, men exit first and pass through doors before women.

12. Japanese are poor at volunteering information. If you want to know something the obligation is on you to ask the right questions. They are well schooled at evasion techniques, while avoiding a positive 'no'. A quick nod means 'I am following you', not 'I agree'. 'Yes' is used in this way: 'You do not want to go, do you?' 'Yes, I do not want to go'.

13. Never assume anything, always check important points again. When you think something important has been agreed in discussion, write a brief memorandum (at the end of each day) and pass it to the Japanese side. If there has been a misunderstanding they will soon let you know and you will be able to discuss the point again.

14. You will appreciate that conducting business in the Japanese way tends to be time-consuming. When you plan your trip to Japan allow yourself more time than you would if you were negotiating business in the western world. You may have to wait for decisions to be made but your time will not be wasted if you use it to gain more experience of the country in a general sense. It may require a succession of visits to establish the right kind of business contacts. Results rarely come quickly in Japan. Successful business is usually the outcome of lengthy, careful and painstaking work in building up the relationship and confidence the Japanese expect. But the Japanese market is big enough to make the rewards well worthwhile for those who are willing to persevere.

Reflections On Relationships

HELMUT MORSBACH

In spite of superficial similarities, as much of this book has demonstrated very forcibly, business in Japan is being conducted on different lines. As a contributor to the last section of this volume, I have been asked to reflect on what has gone before and to give my own opinion as to the continuing success of Japanese business enterprise. Of course a certain amount of 'westernisation' has occurred in the past. But there is rather more to it than that. I believe Japan's success derives from a unique blend of cultural and psychological factors found nowhere else. I would like therefore to highlight what I see as the key points of Japanese perception, thinking and behaviour.

Many factors which influence interpersonal relations are weighted differently in Japan, some of the most important ones being: the closeness between mother and child; the central importance of the group; the importance of reciprocal obligations; the individual's perseverance towards long-range goals; the importance of nonverbal modes of communication.

Of course, these subdivisions are arbitrary and not mutually exclusive. Perhaps they can be viewed as focal points in a complicated web of interrelationships, where one major factor can influence all others and can, in turn, be influenced by them.

The closeness between mother and child

On the whole, training for early independence is an ideal in the West, but not in Japan. Carefully controlled studies have shown that Japanese mothers tend to see their babies much more as extensions of themselves than separate human beings. Physical closeness, including parents and children sleeping together for many years, is regarded positively. This used to be especially so in the case of the eldest son, who traditionally was heir to the house and responsible for his parents in their old age. Dependency was (and still is) highly valued as a warm, life-long feeling which ties one into a chain of superiors (e.g. grandparents and parents) as well as subordinates (e.g. younger siblings). Even acting like a spoilt baby is often seen as positive, and the Japanese word *amae* implies a complex set of emotions and actions which can be described as the *active* desire to be *passively* loved by someone hierarchically superior (originally the mother, later in life the boss, etc.). This does not, however, have negative connotations (common in the West) and also allows the hierarchically superior person, at whom one's *amae* is directed, to 'pet' the subordinate in return. Such behaviour, learned in early childhood, is often carried over into adult life and into the business world.

Western observers might deduce from this that achievement motivation should be low, since in their minds the latter tends to be automatically linked with individualism and the striving for independence. But the reverse is true: Japan is a 'high achievement' society.

The central importance of the group

An understanding of Japanese behaviour is more easily achieved if one views and evaluates individuals above all else in the context of their reference group. In the case of businessmen it is the company they work for and with which they, in time, tend to identify closely. Self-introductions bring this out clearly: Mr Ando, working for the Mitsui Company will introduce himself as, e.g. 'Mitsui's Ando' and talk of his firm or office as '*uchi no. .* ' i.e. 'my home's . . . ' The ideal business is structured like one's family.

In many other countries business can be a reflection of the basic family unit but then it usually involves blood relations and quickly leads to nepotism, low achievement motivation, and inefficiency. It is the great advantage of the Japanese social structure in modern times that family-type organisations can be established on a long-term basis which mainly involve non-kin. Kin are not excluded *per se*, but they do have to prove their individual ability in order to be promoted.

This structure, called *iemoto* (lit.: 'family root') is seen by the anthropologist Francis Hsu as the most distinctive secondary grouping in Japan which has no counterpart in the rest of the world. According to Hsu, its importance lies in the fact that a) it is common in the rural as well as the urban Japan of today; b) it is not restricted to land possession and can thus enlarge and expand country-wide; c) the more complex Japanese society becomes the more objectives it can have; d) it is more of a way of life than just an organisation for its members, who tend to define themselves in terms of their *iemoto*.

Ideally, then, a Japanese, on joining a group, is firmly 'locked into' a set of face-to-face relationships with seniors on the one hand and juniors (in time) on the other. This leads to stable, harmonious work relations and the knowledge that persistent effort will in time be rewarded by movement up the promotion 'escalator'. Therefore it follows that initially getting into the 'right' group (university, company, etc.) is of overriding importance in one's youth — shown by the importance attached to entrance examinations for selection of applicants to desirable groups. Potentially it is eminently democratic because in theory anyone can sit the entrance exams.

It is clear that such a picture presents an ideal which is not always found in practice but the ideal of such a group is very strong and is generally accepted by the Japanese, for it promises stability and harmony inside the group while enhancing competition between groups (e.g. between sections in a firm, between firms, or between countries). For westerners, however, an attempt to be fully accepted into such a tightly-knit system often leads to strong feelings that one's individuality is being slowly smothered. On the other hand, if one has been brought up from childhood to value dependency highly, such group structure provides a warm, accepting atmosphere, as long as one does one's allotted duty and fills one's prescribed role.

The importance of reciprocal obligations

A large amount of personal interaction in Japan (including business) is ritualised. It is not so much the individual personality one seeks when interacting with somebody, but 'proper' role behaviour, i.e. if one is section chief one is expected to behave like one. The Japanese language is eminently suited to this. It has many honorifics (towards the person one addresses, as well as towards somebody of high status one talks about) and many forms of self-deprecation (used of oneself or the group one represents). Japanese reciprocal obligations (weak in the West, because they are ill-defined) can be classed under the headings *on* and *giri*.

On is a feeling of indebtedness which one gets after receiving something needed from somebody hierarchically superior. Its repayment can never be complete, no matter how hard one tries, and brings about feelings (ideally) of life-long loyalty. The origins of the *on*-relationship can be found in experiences of early childhood.

On the other hand, *giri* is a social obligation which requires a Japanese to act as expected by society vis-à-vis another person with whom he has entered into some meaningful relationship. With the aid of *giri*, society can continue to function smoothly, independently of the individual's feelings concerning the order of things or about the other persons with whom interaction must take place. *Giri* explains many 'mysterious' Japanese customs, e.g. the absence of tipping, or the great ritual of gift-exchange in the business world. Furthermore, it throws light on the behavioural dichotomy of *tatemae* and *honne*: *tatemae* is a formal principle acceptable to everyone in order to ensure group harmony, and *honne* refers to one's personal feelings which are rarely divulged. Thus, when westerners accuse Japanese of being two-faced, they are being just that — with the prime concern of not disrupting smooth relations in the group. To westerners, raised on dogmatic Christian thinking, this appears 'inscrutable' or 'false'; to Japanese it is elementary good manners.

Although vertical relations in a group are of great importance, one should not overlook the occasions when equality between members is being stressed, especially at the more informal (but highly important) annual company parties, or after-hours gatherings, where group solidarity is celebrated.

The individual's perseverance towards long-range goals

Although Japanese are strongly group-oriented, there is also a long tradition of *seishin*, i.e. the training of one's own inner mental strength. This is basically an individualistic trait which can only be fully developed by years of practice. It often originated from the respect the child came to have for his long-suffering mother.

Training practices are related to Zen Buddhism, a religious ideal followed, in the past, mainly by the Japanese élite (e.g. *samurai*). However, in recent decades these ideals have filtered downwards and are generally regarded as leading to behaviour which should be emulated. *Seishin* is seen as the opposite of materialism and self-indulgence, allowing one to conquer illness and selfish desires, making stoical acceptance possible of whatever may come.

In essence *seishin* has nothing to do with groups, but it can of course be utilised to further harmonious group interaction. While the flamboyant examples of *seishin* have made world headlines (e.g. the spectacular self-embowelment of the author Yukio Mishima; or the one-man war waged by Lt. Onoda in the Philippine jungle until the mid 1970s) it is more the everyday aspect of perseverance which is so important for an understanding of Japan's industrial strength (like workers who take pride in essentially boring routine tasks, attending to minute details in the running of an organisation where westerners would slacken due to lack of incentive).

The importance of nonverbal communication

Japanese-western relationships are often clouded by the fact that many important aspects of Japanese daily life are rarely, if ever verbalised. Even if verbalising something, there is often a large element of understatement, and silence is valued for its powers of communication. All this is possible only in a highly homogeneous culture (such as Japan) but serious misunderstandings occur in attempts at intercultural communication.

Many of the values mentioned in the previous sections have been transmitted nonverbally. Analyses of motives, and self-doubt, so common in the West, have been absent in large measure in Japan. The isolated location, the history of independence, and the great racial and cultural homogeneity of its people have frequently made words unnecessary where westerners would feel the need for lengthy explanations. While some basic modes of nonverbal communication (nvc), such as smiling when happy, stamping of one's feet in children when angry, etc. can be observed in all cultures, some aspects of Japanese nvc. can be misunderstood by outsiders, who have to learn their real meanings just like one has to learn the spoken language. Some differences include:

a) Body motion. In Japan, the bow as greeting can indicate subtle differences in status and hierarchy, something the handshake is unable to do. Facial expressions can also be different. In Japan, the smile can *also* hide feelings of animosity. Furthermore, there is far less eye contact, since the subordinate rarely looks directly into his superior's eyes. Finally, the 'expressionless' face is an ideal from *samurai* days where one's feelings are hidden from the gaze of one's adversary.

b) In the area of intonation one finds important differences in the tone of voice of the speaker (e.g. being subtly deferential), or the use of silence in speech, which can either be used to convey a powerful message (e.g. in *haragei*, lit. 'stomach art'), or can be passed over without embarrassment as a pleasant period of stillness, which westerners often try to fill with words.

c) The use of space also differs in Japan, e.g. the rigid seating order during official business dinners between two parties. Touching among adults in public is very uncommon; on the other hand, private rooms feel 'crowded' to westerners, although this might be perceived as pleasant by Japanese, who often prefer to sleep together in the same room even if alternative space is available.

d) Finally, the use of artefacts with which one presents oneself to others, is different in Japan. Rank is often indicated nonverbally, e.g. by the uniform-style ways of dressing in the business world, and by the frequent exchange of visiting cards. All these serve to indicate one's rank to others at a glance, making a conversation with the right amount of honorifics, etc. possible.

In spite of surface similarities, The Japanese are not becoming more and more 'westernised'. True enough, periods of extensive borrowing have occurred, especially from China and from the U.S.A. But, when observed closely, these were of an 'osmotic' kind, i.e. the foreign culture was not taken over *en bloc*, but highly selectively. Shaking hands, for instance, did not replace bowing, *kanji* and *kana* are not giving way to the western alphabet.

The Japanese interaction pattern, strange as it may look to westerners on first encounter, is internally consistent to such an extent that it makes Japanese behaviour fairly predictable. But to do that, the westerner must endeavour to become aware of his own unquestioned beliefs and then to try and see things from the Japanese vantage point. However, this is often a difficult thing to do, embedded as we are in a smug way of thinking that western ways are the best and the most efficient.

Selected Bibliography

1 GENERAL: INSIGHTS INTO JAPANESE SOCIETY

AUSTIN L. *Saints and Samurai. The Political Culture of the American and Japanese Elites* (New Haven: Yale University Press, 1975).
Contains pertinent considerations on the role and expectations of managers in regard to themselves and their subordinates.

BENEDICT R. *The Chrysanthemum and the Sword* (London: Routledge & Kegan Paul, 1967).
The work of an American social anthropologist, first published in 1946. A classic sociological study of its kind; well worth exploring.

DOI T. *The Anatomy of Dependence* (Tokyo: Sophia/Kodansha, 1973).
A penetrating and rewarding study of the workings of the Japanese mind and mentality. A best-seller in the original Japanese edition.

GIBNEY F. *Japan: The Fragile Superpower* (New York: Norton, 1975).
Views on Japanese society by a scholar-businessman. Highly informative and perceptive.

GUILLAIN R. *The Japanese Challenge: The Race to the Year 2000* (Philadelphia: J.B. Lippincott, 1970).
Translation from the French (1969). Keen observation based on long years of familiarity with Japan and the Japanese, as well as with East Asia.

HADLEY E.M. *Antitrust in Japan* (Princeton: The University Press, 1970).
Implementation of that policy under the American military occupation and its aftermath. Clear insight into the workings of the large industrial groupings.

HIRSCHMEIER J. AND YUI T. *The Development of Japanese Business 1600-1973* (London: Allen and Unwin, 1975).
Fascinating historical survey providing an historical perspective on current behaviour and practices.

KAPLAN E.T. *The Government-Business Relationship. A Guide for the American Businessman* (Washington, D.C.: U.S. Department of Commerce, 1972).
Most instructive. Especially helpful for three industry studies (steel, automobile, computer) depicting government-business interaction, sometimes a success, sometimes a failure.

MILWARD R.S. *Japan: The Past in the Present* (Tenterden: Paul Norbury Publications, 1979).
A short but lively study of Japan's history, society and commercial structure.

NAKANE C. *Japanese Society* (London: Penguin Books 1973).
Contribution by a foremost Japanese anthropologist. A fascinating account of the vertical structure of Japanese society, its implications and applications.

NORBURY P. (ed.) *Introducing Japan* (Tenterden: Paul Norbury Publications, 1977).
This is a book for the casual reader as well as for those pursuing a specific professional or business interest in Japan as it presents a highly informed view as perceived by a group of outstanding international writers.

OLSON L. *Japan in Postwar Asia* (London: Pall Mall Press, 1970).
Reviews the war reparations agreements and the restoration of commercial relations; then moves into the late sixties when Asia's development came to the foreground.

SCHEINER I. (ed.) *Modern Japan: An Interpretative Anthology* (New York: Macmillan, 1974).
Provides also an historical perspective, and puts special stress on thought patterns.

SHIBA K. *Oh, Japan! Yesterday, Today & Probably Tomorrow* (Tenterden: Paul Norbury Publications, 1979).
A witty, informed and often very instructive analysis by one of Japan's veteran and greatly admired journalists.

Staff of *Asahi Shimbun. The Pacific Rivals: A Japanese View of Japanese-American Relations* (New York and Tokyo: Weatherhill, 1972).
Collection of short essays by the staff of Japan's largest newspaper. A best-seller in Japanese. Interesting reading taking the pulse of post-war Japanese public opinion as moulded and expressed by the press.

VOGEL E.F. *Japan's New Middle Class: The Salary Man and His Family in a Tokyo Suburb* (Berkeley: University of California Press, 1963).
   Although depicting the situation at the end of the 1950s, this sociological study remains a major contribution to the understanding of the current consumer market and its behaviour patterns.

YANAGA C. *Big Business in Japanese Politics* (New Haven: Yale University Press, 1968).
   Excellent contribution to the understanding of the government-business relationship from the viewpoint of a political scientist.

YOSHIDA K. *Japan is a Circle* (Tenterden: Paul Norbury Publications, 1975).
   An entertaining, satirical and informative tour in essay style.

2 BUSINESS KNOW-HOW

ABEGGLEN J.C. *Management and Worker — The Japanese Solution* (Tokyo: Sophia/Kodansha, 1973).
   Reprint of the classic *The Japanese Factory* (1958), to which was added a comprehensive analysis of the employment system in the 1970s.

BALLON R.J. (ed.) *Marketing in Japan* (Tokyo: Sophia/Kodansha,1973).
   The topic is treated under three central headings: setting-up in Japan; the Japanese consumer; marketing strategy for Japan.

BALLON R.J. and LEE E.H. (eds.) *Foreign Investment and Japan* (Tokyo: Sophia/Kodansha, 1972).
   Review of the investment situation in Japan and its critical areas: corporate control; antimonopoly legislation; taxation; patents and trademarks; the management of foreign operations.

BALLON R.J., TOMITA I. and USAMI H. *Financial Reporting in Japan* (Tokyo: Kodansha International, 1976).
   The current financial system and corporate financing; the problems of disclosure and analysis of financial reporting.

BIEDA K. *The Structure and Operation of the Japanese Economy* (Sydney: Wiley, 1970).
   Overall review. Provides detailed insights into many lesser known practices. Via the index, a helpful source of reference.

BLUMENTHAL T. *Saving in Postwar Japan* (Cambridge, Mass.: Harvard University Press, 1970).
Lucid analysis of one of the most intriguing aspects of Japanese society — the capacity to save.

Case Examples of Wage and Labor Regulations (Tokyo: Business Intercommunications, 1977), 2 vols.
English and Japanese; all cases relate to foreign operations in Japan.

DOI T. *Digest of Japanese Court Decisions in Trademarks and Unfair Competition Cases* (Tokyo: American Chamber of Commerce in Japan, 1971).
An analysis of thirty-two recent cases. Appendix includes the Trademark Law and the Unfair Competition Prevention Law.

DOI T. and SHATTUCK W.L. (eds.) *Patent and Know-how Licensing in Japan and in the United States* (Seattle: University of Washington Press, 1977).
Comparative study of the laws, Patent Office, court procedures, agreements, antitrust problems and taxation.

DORE R. *British Factory — Japanese Factory. The Origins of National Diversity in Industrial Relations* (London: George Allen & Unwin, 1973).
Lengthy but extremely helpful comparative analysis. Via the index, an enlightening reference source on almost all aspects of personnel administration and industrial relations.

HENDERSON D.F. *Foreign Enterprise in Japan: Laws and Policies* (Chapel Hill: University of North Carolina Press, 1973).
Comprehensive study of the investment climate in present day Japan. Probes the depths of most problems encountered by foreign firms. Written by a lawyer.

FRANK I. (ed.) *The Japanese Economy in International Perspective* (Baltimore: John Hopkins University Press, 1975).
Economic and industrial policies; distribution; foreign trade and foreign direct investment.

The Industrial Policy of Japan (Paris: OECD, 1972)
Incisive analysis of a broad topic. Based on extensive documentation provided by the Japanese government.

Manual of Employment Practices in Japan, revised edn (Tokyo: American Chamber of Commerce in Japan, 1970).

Meant for the resident foreign executive confronted with the problem of managing a Japanese work-force. Periodically updated.

MARSHALL B.K. *Capitalism and Nationalism in Prewar Japan — the Ideology of the Business Elite, 1869-1941* (Stanford University Press, 1967).
Illuminating study. Will help in the understanding of the mentality of present-day Japanese executives in their fifties.

Outline of Japanese Distribution Structures (Tokyo: Distribution Economics Institute of Japan, 1973).
Indispensable guide. General considerations on the distribution environment and structure, followed by detailed analysis of the practices and channels for each major category of consumer goods. Illustrated by diagrams.

ROHLEN T.P. *For Harmony and Strength: Japanese White-collar Organization in Anthropological Perspective* (Berkeley: University of California Press, 1974).
Detailed survey of personnel administration in a local bank.

TANAKA H. (ed.) *The Japanese Legal System. Introductory Cases and Materials* (Tokyo: University Press, 1976).
Of particular interest to the businessman are the views on the role of law, judges and lawyers in Japanese society.

TSURUMI Y. *The Japanese Are Coming. A Multinational Interaction of Firms and Politics* (Cambridge, Mass.: Ballinger, 1976).
The rise of Japanese investments overseas and the role of trading companies and banks; transfer of technology; ownership and image.

TSURUMI Y. *Japanese Business. A Research Guide with Annotated Bibliography* (New York: Praeger, 1978).
Helpful bibliography to journal articles as well.

VAN HELVOORT E. *The Japanese Working Man. What Choice? What Reward?* (Tenterden: Paul Norbury Publications, 1979).
For the author, Japan's success lies to a very great extent in the human factor as demonstrated within the remarkable personnel management system.

YAMASAKI Y. *Digest of Japanese Court Decisions in Patentability and Patent Infringement Cases, 1966-1968* (Tokyo: American Chamber of Commerce in Japan, 1970).
This is the third volume on the subject (available from the same

publisher). Depicts the evolution in judicial thinking through a three- to four-page digest of forty cases. Index to the three volumes.

YOSHINO M.Y. *Japan's Managerial System — Tradition and Innovation* (Cambridge, Mass.: MIT Press, 1968).
In-depth analysis starting with the pre-modern and prewar eras; considers present-day Japanese executives and their ideologies, the industrial structure, industrial groupings, the organisation structure and decision-making.

YOSHINO M.Y. *Japan's Multinational Enterprises* (Cambridge, Mass.: Harvard University Press, 1976).
The procurement of raw materials; the spread of manufacturing overseas; the role of the trading company.

YOSHINO M.Y. *The Japanese Marketing System — Adaptations and Innovations* (Cambridge, Mass.: MIT Press, 1971).
Exhaustive study covering the postwar emergence of the mass-consumption society; the marketing behaviour of the large manufacturing firms; tradition, innovation and government policies in the distribution sector; and consumer financing.

A Wage Survey of Foreign Capital Affiliated Enterprises in Japan 1978 (Tokyo: Business Intercommunications, 1978).
Their wage systems and examples from each particular industry.

3 PERIODICALS/ANNUALS

Economic Plan for the Second Half of the 1970s: Toward a Stable Society (Tokyo: Economic Planning Agency, 1976).

The seventh post-war economic plan of the Japanese government.
Economic Survey of Japan (1977-1978) (Tokyo: The Japan Times, 1978).
Translation of the annual survey presented by the Economic Planning Agency to the government. Besides key economic data, presents the official thinking with regard to economic, social and international implications.

Economic Surveys: Japan (Paris: OECD, 1978).
Annual authoritative survey at the aggregate level.

Japan Economic Yearbook 1978-1979 (Tokyo: The Oriental Economist, 1978).
Annual following the universal pattern of such yearbooks.

Detailed current data on society in general and on each industry, as well as a list of major companies classified by industry. Useful reference.

Quality of the Environment in Japan (Tokyo: Environment Agency, 1977).
Condensed version of the annual White Paper on the Environment. Necessary source for the exporter of pollution-control equipment and related industries.

White Paper on International Trade 1978 (Tokyo: Japan External Trade Organisation, 1978).
An annual analysis of Japan's international trade. Detailed statistics by region, country and commodity, supported by comments on the international environment viewed from the Japanese angle.

White Papers of Japan, 1976-1977. Annual Abstract of Official Reports and Statistics of the Japanese Government (Tokyo: Japan Institute of International Affairs, 1978).
Annual publication containing about ten-page abstracts of the White Papers published by various government agencies, as well as public opinion surveys, an outline of the political system and detailed organisation charts of the Japanese government agencies.

List of Contributors

ROBERT J. BALLON. Born in Belgium but a resident of Japan since 1948. Has concentrated on the study of Japanese labour-management relations and Japan's role in international business. His books include *Doing Business in Japan, Joint Ventures in Japan, The Japanese Employee, Japan's Market & Foreign Business, Foreign Investment and Japan, Marketing in Japan* and *Financial Reporting In Japan*. He is Chairman of the Socio-Economic Institute at Sophia University, Tokyo, and head of the International Management Development Seminars sponsored by his institute.

GREGORY CLARK. A former Chinese and Soviet Affairs specialist in the Australian Foreign Ministry, serving in Hong Kong and Moscow in the early 1960s. He resigned the service in 1965 and after four years of post-graduate research into Japanese direct investment abroad, he became Far East Correspondent for *The Australian*. He continues to live in Japan, writing and lecturing. In 1977 he published in Japanese his own views on the nature of Japanese society which quickly became a best-seller.

GEORGE FIELDS. An Australian citizen born in Japan. After secondary education at a Japanese high school he obtained an Honours degree in Economics from the University of Sydney. He is today Chairman and Chief Executive of ASI — one of the largest western origin consumer research companies in Japan. He lectures regularly in both English and Japanese.

DAVID GRIBBIN. Has spent over ten years living in Japan and working closely with the Japanese advertising industry. Today he is a senior consultant to one of Japan's leading advertising agencies and is a regular visitor to Europe acting on behalf of major clients.

GENE GREGORY. A specialist in Asian affairs via a career including diplomacy, business and journalism. His home is Geneva but he is a regular visitor to Japan and is currently a visiting lecturer at Sophia University, Tokyo. He is a regular contributor to a number of international publications.

SIMON GROVE. For several years chief coordinator in Tokyo for a major London broking firm and today the Japan Representative for the same company with a permanent office and staff which he organised and established. As an international business consultant, his interests are wide-ranging and include books and publishing.

MASAAKI IMAI. President of the Cambridge Corporation, Tokyo, a major international recruitment firm which has assisted over 4000 companies both foreign and Japanese. He is the author of several books on business in Japan, including *Change Jobs Successfully* and *The Japanese Businessman*.

MICHAEL ISHERWOOD. Manager, General Affairs Dept. and Assistant to the Director, Mitsubishi Corporation, London. In nearly twenty years with Mitsubishi he has had a wide range of experience in different trading activities between Japan and Europe. He holds directorships in several subsidiary companies and is a regular speaker at business seminars.

SHINZO KATADA. Has worked for Nomura Research Institute for twelve years and has spent the last two in the London office as Senior Industrial Economist. His specialist research includes the automobile industry and Japanese companies operating in S. E. Asia. In London he has been mainly engaged in introducing the developments of the Japanese economy and industry to Europeans investing in Japan.

KUNIHIKO KOBAYASHI and TAKASHI SUGIYAMA. Both contributors are currently Chief Manager and Deputy Chief Manager respectively of Kojimachi Branch, Mitsubishi Bank, Japan. Mr Kobayashi brings wide international experience to his branch which is a Tokyo district well known for its high percentage of foreigners and foreign business interests.

JOHN KIRBY. Is adviser for Asian Affairs in the Bank of England. He visits Japan regularly and was Financial Attaché in the British Embassy in Tokyo in 1974–76.

HELMUT MORSBACH. Senior Lecturer in Psychology at Glasgow University. He has been a frequent visitor to Japan and has lived there for extended periods. During the summers of 1977 and 1978 he was Visiting Research Fellow in the Psychology Department of the International Christian University, Tokyo. He has published widely on Japan and won the 1978 essay award from the International Cultural Association of Kyoto.

TERUYASU MURAKAMI. Has worked for Nomura Research Institute for twelve years and has spent the last three years in the London office as Senior Consultant. His specialist research includes the behaviour and strategy of the multi-national corporation and the systems approach to economic and social development of developing countries as well as the OPEC economies.

MARTYN NAYLOR, MBE. After graduating from the London School of Economics in 1957, he received an MBA from the University of New Mexico where he also taught. In 1960–61 he lectured in economics at the University of Hong Kong. He has lived in Japan since 1965 and today is president of Naylor Hara International K.K., his own marketing and advertising company. He has been Honorary Secretary of the Japan-British Association since 1968.

SADAO OBA. After graduating from Otaru College of Commerce he spent 32 years working for Mitsui & Co., including ten years in London responsible for general economic surveys and external relations. In 1979 he joined Nomura Research Institute as Senior Consultant. He is the author of several publications and is a regular speaker at business conferences and seminars.

MASAO OKAMOTO. Elected to the board of Nomura Research Institute, Japan, in 1975 and is currently Director of their International Studies Department. A graduate of Osaka City University of Commerce, he earlier spent 16 years with JETRO (Japan External Trade Organization) when he served in Lagos, Brussels and London. During this period he wrote five of the Overseas Market White Papers.

JONATHAN RICE. Director and General Manager, Smiths Industries Japan Ltd. with extensive experience of living and working in Japan.

JOHN ROBINSON. Deputy Manager, Barclays Bank International Ltd., Tokyo.

CHARLES SMITH. Far East Editor, *Financial Times*, Tokyo, a position he took up in 1973.

SUMIO TAKEUCHI. Head of his own law office, Tokyo, with wide experience of Japan-West litigation and legal affairs generally. He lectures regularly and sits on various advisory committees. He graduated from the Faculty of Law, University of Tokyo, in 1959 and later studied at various American universities while working for law firms in New York and Chicago.

ANDREW WATT. Since 1976, General Manager, Japan Market Research Bureau — a division of the J. Walter Thompson advertising company. He is a Council member of the Japan-British Society and a board member of the Asiatic Society of Japan. His publications include *The Truth About Japan* and *Marketing in Japan.*

THE EDITORS

PAUL NORBURY is Managing Director of Paul Norbury Publications Ltd.

GEOFFREY BOWNAS is Professor of Japanese Studies and Director of the Centre of Japanese Studies, University of Sheffield.

Anatomy of a major Japanese trading company

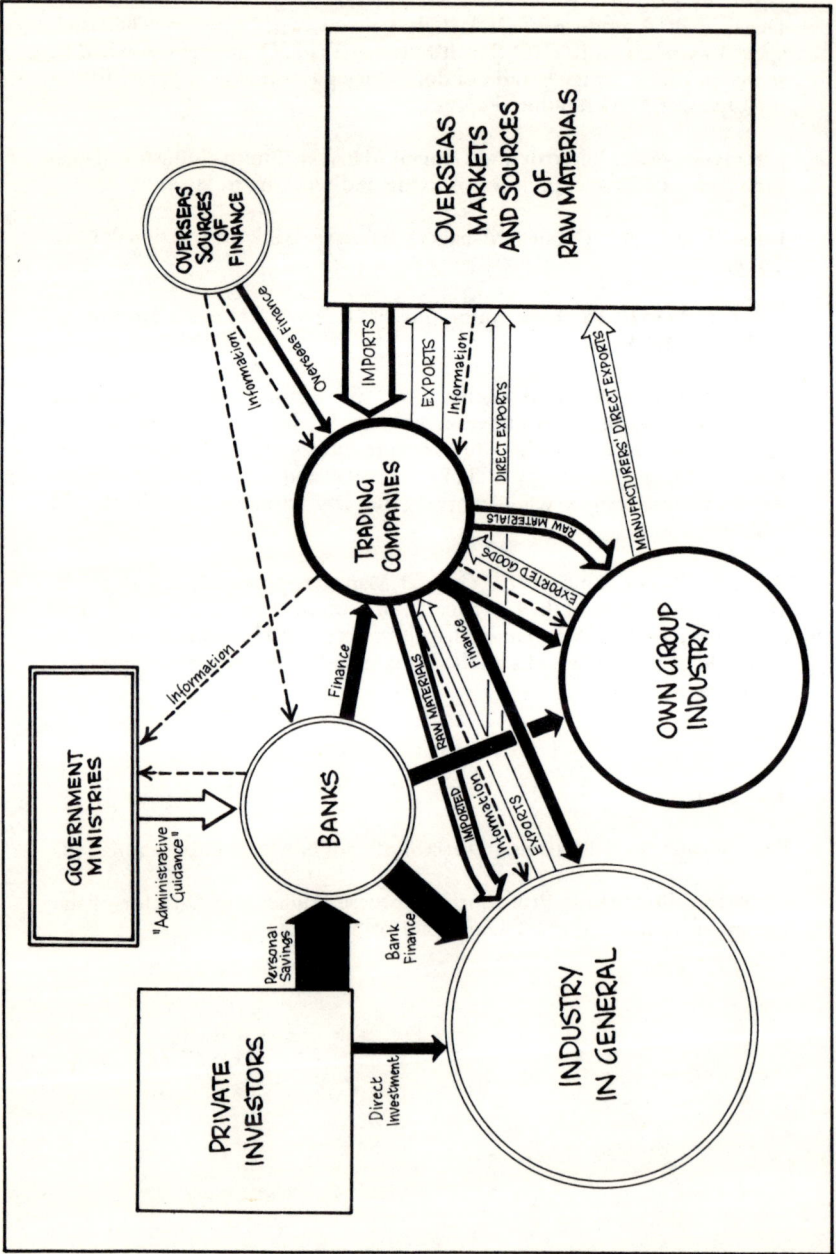

Index

Accounting and auditing, 152ff
Administrative guidance, 120, 159 ff
Advertising, 40, 68-76, 151
Adviser, 167
Air-conditioning, 16
Air transport, 14
Algemene Bank, 97
All Nippon Airways, 14
Amae, 5, 191
American Express International Banking Corporation, 97
Asahi, 113
ASEAN, 22
Ashwell K.H., 31
Ataka & Co., 131
Automobile industry, 13, 18, 43
Avon, 36

Bank of America, 96
Bank of Japan, 89, 101, 114, 119, 145, 147, 163
Bank of Tokyo, 97
Banking system, 85-102, 110, 114, 118
Barclays Bank International Ltd., 97
BASF Japan, 37
Board of Directors, 146
Bocuse P., 74
Bonus system, 42, 79, 111, 164, 167
Bossers C., 33
Budhism, 73, 111, 193
Braun, 33
Bruynzeel, 32
BSR, 35
Bucho, 126, 129, 178
Business career, 128-30
Business interrelationships, 109-15

Campari, 74
Canon, 13
Capitalisation, 118ff
Cards, business, visiting, 178, 187, 195
Cash cards, 100
Cash dispenser, 100
Central Finance, 11
Chase Manhattan Bank, 96, 97
Chuo Bussan, 31
Citibank, 96
C. Itoh, 131, 132

Coca-Cola, 30, 71
Clough A.H., 67
Commercial code, 118, 141, 144, 146, 152, 153, 156, 184
Company President, 124ff
Company structure, 110ff (*see also* Management)
Computers, 12
Confucianism, 111
Construction, 17
Consumer credit, 13ff, 91
Consumer market, 41ff
Consumer patterns, 40ff, 76ff, 91
Contracts, 143ff
Copel Inc., (U.S.A.), 133
Copying machines, 13
Corporate finance, 90
Corporate Reorganisation Law, 157
Corporation Law, 156
Culture, 1-9, 46, 76ff, 109ff, 190-5
Custom Office Exclusion Procedure, 158

Daiei, 15, 36, 41
Daiichi Kangyo Bank, 113
Daishinpan, 11
Daiwa, 104
Dendenkosha, 171
Dentsu, 72
Department Store Law, 58
Department stores, 15, 33, 58, 61
Distribution system, 30, 57-63, 70, 151
Downham J., 65

Economic environment, 85-94
Economic relations with other States, 21ff
Education, 44, 49, 111, 114
Eigo-ya, 75
Electric appliances, 18-20
Electrical construction, 16
Electronics, 19, 26
Employment Security Office, 123
Employment system, 7, 117ff, 148, 164ff, 184ff
Entertainment, 178ff, 188, 194
Estée Lauder, 31
Etiquette, 177-89

Exports, to East Asia, 22ff; to U.S. and Europe, 29

Fair Trade Commission, 145, 151
Family ties, 7, 78ff, 190ff
Fashion, 14, 48, 52
Fertilisers, 17
Financial system, 86ff, 94ff
Financial Statements, 153ff
Food, 15, 16, 43, 51, 76
Foreign banks, 94
Foreign exchange, 96
Foreign Investment Council, 145
Fuji Bank, 113
Fujitsu, 12
Fumio Yamamoto, 74

GAAP (Generally Accepted Accounting Principles), 152
G.F., 62
Gensaki, 89
Giri, 192
Giri-ninjo, 5
Government agencies and activities, 145
Government bond market, 88
Group Society, 1-9, 109ff, 113, 117ff, 136, 148, 184, 191
Gruner & Jahr, 68

Hattori, 173
Hattori Tokeiten, 62
Health, 12, 43, 50, 111
Heavy machinery, 18
Heublin (U.S.A.), 133
Hideyuki Oka, 71
Hitachi, 11, 12, 106
Holiday Inns Far East Ltd, 132
Honda, 32
Hong Kong, 22ff
Honne, 65, 181ff
Household expenditure, 42ff
Housing, 1, 53, 111
Housing Lease Law, 170
Hsu F., 191

IBM, 12
Imports, from East Asia, 24ff; from Europe and U.S.A., 37
Individual rights, 161
Indonesia, 22
Industry, 10-28, 85ff
Interest rates, 88
Internationalisation of yen, 92
Iran-Japan Petrochemical Corporation, 133
Iranian National Oil Corporation, 132

Ito-Yokado, 15

The Japan Company Handbook, 108
The Japan Economic Journal, 108
Japan Kentucky Fried Chicken, 133
Japan National Railway, 14, 16
Japanese banks, 98-102
JETRO, 136, 163
Johnson & Johnson, 31, 33
Joint Ventures, 99, 122, 137-55
Jomukai, 124
Jung H.F., 30
Jusco, 15

Kabushiki kaisha (KK), 106, 147, 156ff
Kacho, 126, 128, 178
Kaicho, 146
Kaisha Shikiho, 108
KDD, 171
Keidanren, 37, 121
Keizai Doyukai, 121
Kirin, 71, 113
Kokusai Denwa Denshin Kyoku, 171
Konishiroku Photo Industry, 13

Landlord, 169
Language problems, 2, 30, 108, 164, 185, 187, 194
Large Retailers Law, 15, 59
Law, 15, 143ff, 155-62; Corporation Law, 156; Rule of law, 160
Leasing companies, 11
Leisure, 46ff, 80
Leyland Japan, 132
Life, 11
Lipton, 31
Lockwood, W.W., 121

MacDonald, 40
McVities, 71
Malaysia, 22
Management structure, 115-30, 185ff
Manufacturing industries, 18
Market research, 62, 63-7
Marketing, 29-84, 150
Marks & Spencer, 36, 41
Marriage, 137
Marui, 91
Matsubo Electronic Components Sales Co., 32
Matsubo Electronic Instrument Sales Co., 32
Matsushita, 32, 69, 74
Medical equipment, 11
Meiji, 71
Meiji Restoration, 116, 155, 185

Melitta, 31, 35
Ministry of Construction, 122
Ministry of Finance, 87ff, 114, 122, 152, 158, 163
Ministry of Health and Welfare, 122
MITI, 12, 120, 139, 154
Mitsubishi, 62, 71, 113, 123, 133
Mitsubishi Bank, 98, 101, 113
Mitsubishi Estate Company, 113
Mitsui, 31, 113, 123, 133, 134
Morgan Guarantee Bank, 97
Moriguchi Isao, 82
Morinaga, 32

Nescafé, 71
Nestlé 31
Nichii, 15
Nihon Keizai Shimbun, 108, 178
Nihon Philips, 31
Nikko, 104
Nikon, 113
Nippon Telegraph and Telephone Public Corporation (NTT), 19, 20, 171
Nippon Electric, 12
Nippon Shinpan, 11
Nissan, 13
NTT, 20, 171
Nomura Research Institute, 10
NYK, 113

Office, setting up, 163-72, 176
Office location, 168
Office staff, 163-8
Ogilvy, D., 74
Okita Saburo, 25
Olivetti, 31
On, 192
Overdrafts, 101

Petrochemicals, 17,18
Pez, 32
Philips, 33, 62, 70
Philips Industrial Development and Consultants Co. Ltd., 33
Pollution, 8
Process industries, 19
Processed meat, 16
Public transport, 43

R & D, 149
Roppo Zensho (*The Six Codes Book*), 155
Retail Control Special Measure Law of 1959, 58
Retirement, 111
Ricoh, 13
Road building, 17

Salaries, 167
Samurai, 116, 194
Sanwa Bank, 113
Savings, 104, 110
Sears Roebuck, 92
Securities, 102ff
Security Exchange Act, 153
Seishin 193
Service industry, 9, 33, 40, 44
Sharp, 13
Shasho, 146
Shipbuilding, 17, 18, 25
Singapore, 22ff
Smiths Industries, 173-6
Soft drinks, 15
Sogo shosha, 23, 131ff (*see also* Trading companies)
Sony, 32, 82
Sony Plaza Co., 32
South Korea, 22ff
Steel, 17, 18
Stock exchange, 102-8
Sumitomo, 113
Suntory, 71
Superstores, 15, 58

Taces, 11
Taiwan, 22ff
Tampax, 31
Tanaka Government, 86
Tannoy, 36
Tatemae, 65, 181ff
Technics, 74
Telecommunications, 19ff
Texas Instruments, 19
Textiles, 17, 25
Thailand, 22
Tokugawa Administration, 73, 110, 120, 184
The Tokyo Weekender, 108
Toshiba, 106
Toshiba Credit, 11
Toshoku, 31
Tourism, 38, 47
Toyoda Tsusho Kaisha Ltd., 32
Toyota, 13, 32
Trade, in East Asia 21ff; in Europe and U.S.A., 37
Trademark Law, 158
Trade unions, 117ff
Trading companies, 61, 112ff, 131-6
Twining, 40

Union Bank of Switzerland, 97

Wage structure, 117

Wakon yosai, 116
Warner-Lambert, 62
Watches, 14, 38, 173
Westdeutsche Landesbank, 97
Whirlpool, 32
Wholesalers, 59ff, 150

Xerox, 13

Yamaichi, 104
Yoshida, 92
Yukichi Fukuzawa, 119
Yukio Michima, 193
Yugen kaisha, 147, 156ff